Praise for *Goats in America: A Cultural History*

"Truly, I thought I knew a lot about goats until I read Tami Parr's *Goats in America*! I was blown away by the depth of information and delightful stories contained in this book. This book will appeal not only to goat lovers, but to anyone who enjoys exploring history from a unique perspective."

—Gianaclis Caldwell, author of *Holistic Goat Care: A Comprehensive Guide to Raising Healthy Animals, Preventing Common Ailments, and Troubleshooting Problems*

"Who knew goats would make such a rich and captivating topic? Fortunately, Tami Parr did, and her exhaustive research and lively writing have produced a riveting narrative. Goats have had a turbulent trajectory in the United States, banished in some eras, beloved in others. Thanks to Parr, I have renewed appreciation for these gentle creatures that provide us with such nutritious milk, meat, and cheese."

—Janet Fletcher, author and publisher of the *Planet Cheese* blog

GOATS
in AMERICA

A Cultural History

TAMI PARR

Oregon State University Press   Corvallis

Oregon State University Press in Corvallis, Oregon, is located within the traditional homelands of the Marys River or Ampinefu Band of Kalapuya. Following the Willamette Valley Treaty of 1855, Kalapuya people were forcibly removed to reservations in Western Oregon. Today, living descendants of these people are a part of the Confederated Tribes of Grand Ronde Community of Oregon (grandronde.org) and the Confederated Tribes of the Siletz Indians (ctsi.nsn.us).

Cataloging-in-Publication data is available from the Library of Congress (LCCN 2025019764)

ISBN 978-1-962645-45-4 paper; ISBN 978-1-962645-46-1 ebook

∞ This paper meets the requirements of ANSI/NISO Z39.48–1992 (Permanence of Paper).

© 2025 Tami Parr
All rights reserved.
First published in 2025 by Oregon State University Press
Printed in the United States of America

COVER PHOTO: Pearl from Villa Villekulla Farm in Bethel, Vermont. All photos by the author unless otherwise noted.

Oregon State University
OSU Press

Oregon State University Press
121 The Valley Library
Corvallis OR 97331–4501
541-737-3166 • fax 541-737-3170
www.osupress.oregonstate.edu

Contents

Introduction   1

Part I: Poor Man's Cow
Chapter 1: Poor Man's Cow   9
Chapter 2: Navajo Goats: Herding and Administrative Control   29

Part II: Commercial Success with Dairy Goats
Chapter 3: How Goat's Milk Became Healthy   51
Chapter 4: The Goat's Milk Business   71
Chapter 5: Back to the Land: A Dairy Goat Renaissance   93
Chapter 6: Say Chevre   119

Part III: Contemporary Goat Culture
Chapter 7: Urban Goats   141
Chapter 8: Goat Meat in America   161

Acknowledgments   185
Notes   187
Selected Bibliography   213
Index   217

Introduction

One morning in the spring of 2023, a group of about thirty-five goats were spotted running loose through the streets of San Francisco. Social media feeds spread images of the herd running toward Fisherman's Wharf with scenic San Francisco Bay and Alcatraz Island in the background. According to reports, the goats paused at a stop sign as they made their way down a series of steep hills.

Prior to their adventure, the renegade goats were employed grazing the grassy landscape at Francisco Park in the city's Russian Hill neighborhood, charged with clearing poison oak and other invasive plants. Representatives from City Grazing, a Bay Area nonprofit organization that deploys goats for grazing purposes across the region, said that a portion of fencing had been vandalized, leaving an opening that enabled the goats to escape. The incident was short-lived: within about a half an hour, police and neighborhood residents rounded up the goats, enticing them with a bale of hay, and the herd was returned to Francisco Park. Though the moment caused a certain amount of stress for San Francisco police and employees of the city's Parks and Recreation Department (not to mention the folks at City Grazing), overall the goat escapade inspired joy and delight in onlookers—although one neighborhood resident was reported to have complained about goat poop on the sidewalk.

While the runaway goats made for an entertaining feature on the evening news, this was actually far from first time that goats have roamed through the streets of San Francisco. During the nineteenth and early twentieth centuries, wandering goats were a common sight in cities across the country, from San Francisco to Chicago to New York City. Back then reactions to the animals were far more negative, however. Residents chastised municipal officials for their inability to rid the city of nuisance goats running amok.

This book examines the history and culture of goats in the United States. For thousands of years, the domestic goat has proven quite useful to humans, providing sustenance in the form of meat and milk along with other valuable products, including hides for leather and fiber for textiles. Yet despite their utility, for most of American history goats were considered expendable—useful if necessary, but avoided if possible. Over the centuries, however, Americans have changed their collective minds about goats. In contemporary society, city residents keep goats in their backyards. Goats are regularly featured in movies and TV commercials. And these days, if people see goats running through city streets, they are more likely to laugh delightedly while photographing or recording the event than to call the authorities. How did we get here? This is the central question I will explore throughout the course of this book.

Domesticated goats first arrived on the shores of North America along with other livestock like cattle, pigs, and horses starting in the sixteenth century. Spanish conquistadores introduced goats to the Caribbean islands, Mexico, and eventually to what is now the Southwestern United States. Meanwhile, goats and other livestock sailed on the ships of English, Dutch, and French colonists to the Atlantic Coast of North America. As European colonies expanded on both sides of the continent, so too did the goat population.

Livestock were central to the survival of European colonists. Cattle, for example, provided meat and dairy products, as well as labor in the form of plowing and hauling. At the same time, goats were smaller and less profitable than cattle and did not produce wool like sheep. Goats were also more difficult to manage than other types of livestock and became despised for destroying crops and consuming fruit trees. During the seventeenth century, many colonial municipalities on the East Coast banned goats entirely.

After the Revolutionary War, the United States expanded rapidly and the Industrial Revolution upended the agrarian landscape and economy. Immigrants from Europe streamed to America to work in factories in fast-growing cities. The poor and working classes in industrializing cities across the country often kept goats for sustenance and survival. The animals became the scourge of urban life, running recklessly through the streets, destroying property, and sometimes even attacking humans in their search for food. Goats became the punch line of jokes and potent symbols of degenerate behavior of all kinds.

By the nineteenth century, entrenched negative perceptions of goats began to shift amidst the milk contamination scandals of the era. Consumers were shocked to learn that the cow's milk they were drinking was often adulterated with substances like chalk or plaster of paris and laden with dangerous microorganisms, including the bacterium that causes tuberculosis. Many turned to goat's milk as a substitute, and subsequently an industry evolved to meet growing national demand. Enterprising farmers with dollar signs in their eyes began importing productive European dairy goat breeds such as Toggenburgs and Saanens from Europe to the United States. By the turn of the twentieth century, goats were celebrated as saviors.

After World War II, farmers and back-to-the-landers across the United States further reinvigorated the practice of goat keeping. Goats were the perfect animal for those interested in healthy food and self-sufficiency but unfamiliar with livestock farming. As a result, goat's milk became a popular health food staple and during the 1960s and '70s, small goat farmers like Laura Chenel began to produce and sell goat's milk cheeses to an increasingly food curious public. Goat's milk cheese production has since become a multibillion-dollar industry in the United States. Contemporary livestock farmers are more likely to see goats as a profit center than a nuisance.

Even if you're not partial to goat's milk cheese, you may have encountered a goat recently—perhaps in someone's backyard now that urban goats are all the rage. Maybe you have attended a birthday party where baby goats were available to be petted and photographed. Goat memes are all over social media, as is the acronym GOAT (greatest of all time), a term used to refer to someone possessing exceptional abilities. Media outlets have proclaimed that goats are the "new dogs." In the twenty-first century, goats are not just popular, they're inescapable.

I first became interested in goats in the course of writing my two previous books, *Artisan Cheese of the Pacific Northwest* and *Pacific Northwest Cheese: A History*. While immersed in the world of cheese and cheesemakers around the region, I was lucky enough to visit many goat farmers and goat cheesemakers. I also met a lot of beautiful Alpine, Saanen, and Nubian dairy goats who would not be ignored. Once I started researching the animals, it became clear that the history and social significance of goats in our culture has been mostly overlooked. While mountains of material have been written over the decades about other types of livestock, particularly cattle and the development of the American meat industry, comparatively little attention has been directed toward goats. I think it's fair to say we don't understand goats at all.

I hope this book inspires curious people to start thinking more seriously about goats and the diverse and important roles the animals have played throughout the history of the United States. While I've touched on key moments and significant developments in goat history and culture in this book, there is so much more to be uncovered.

For hundreds of years, Americans have used goats to their own advantage, but they have also projected their desires, fears, and dreams onto the animals. Goats have been cast as destructive and evil, symbols of social degradation, and more recently as impossibly cute and cuddly. But in the final analysis, the history and reputation of goats in American society really says more about we humans than about goats themselves. Whatever any of us thinks about goats (or goat meat or goat yoga), our opinions are not really reflective of the qualities or capabilities of goats as an animal or a species. Goats are hardy, curious ruminants who have supported and endured humans across the globe for centuries. No doubt they will continue to do so for many more to come.

Souvenir Goldie the Goat toy from Simone Biles's 2021 Gold over America Tour. Biles has regularly used goat imagery as a play on the phrase "Greatest Of All Time." Photo by Kim Hogeland.

I

Poor Man's Cow

CHAPTER ONE

Poor Man's Cow

> *To Mayor Chapin of Brooklyn:*
> *Can't you prevail on your party constituents of Brooklyn this Summer to keep their goats out of our yards, where they destroy flowers and gardens, and also off the streets where they destroy shade trees? Let the police shoot them and cart them away—the goats I mean.*
>
> Alice Sherman, 1890

The story of goats in America begins in Europe in the fifteenth century. That's because goats and other familiar domesticated livestock like cattle and sheep are not native species to the Western Hemisphere. Goats and a variety of other animals first arrived in the Caribbean islands on the ships of Christopher Columbus's second voyage in 1493. Later those Spanish goats spread throughout present-day Mexico and into what would eventually become the Southwestern United States. Meanwhile, European nations, including England, France, and Holland, brought their own goats, horses, cattle, and other animals to America during the seventeenth century as they colonized the Atlantic Coast of North America. Soon enough, European livestock began to multiply and proliferate across what we now know as North and South America.

Goats were an important component of the business of global oceanic travel and colonization. Ships of the era were very small, and much of their limited capacity was taken up by food and supplies necessary for extended ocean journeys. Because of their size, goats often accompanied these voyages and were employed as a mobile milk source. In times of need, sailors could also slaughter a goat for meat. In addition, European explorers often left goats and other animals, including pigs, on remote islands all over the globe, with the idea that later ships might come along and use the animals or their descendants for food—the islands serving essentially as maritime highway rest stops. Spanish explorer Juan Fernández planted goats and other livestock on the islands now named after him off the coast of Chile;

Portuguese explorers left goats on the Cape Verde Islands and on Saint Helena in the South Atlantic, to name just a few examples. Descendants of some of those transplanted island goats are still around today; the Arapawa goats of New Zealand, for example, are thought to be descended from goats first placed in the region by English Captain James Cook during the eighteenth century.

Probably the most celebrated of the exploration era's maritime caprines was an English goat (name unknown) that traveled around the world twice, once with Captain Samuel Wallis and a second time with Captain James Cook. Upon her retirement from service, she was said to have received her own garden plot for grazing at Captain Cook's home in London. A collar around her neck bore an inscription penned by Dr. Samuel Johnson:

> In fame scarce second to the nurse of Jove
> This goat, who twice the world has traversed round,
> Deserving both her master's care and love,
> Ease and perpetual pasture has now found.

The goat's obituary appeared in a London periodical in 1772.[1]

❖

Andries Jochemsen was a sail maker and tavern keeper in seventeenth-century New Amsterdam, a Dutch colony on the Atlantic Coast of North America. Like countless landlords before and since, Jochemsen had a problem tenant. In a suit filed against the tenant, Jochemsen accused lessee Claes Claesen Smitt of allowing the Jochemsen property, specifically the apple trees, to be destroyed by goats. Smitt countered that he'd been conscripted into service, possibly by the Dutch West India Company, and therefore wasn't responsible for any property damage that occurred in his absence.

In a similar case brought before the courts of New Amsterdam, Willem Bredenbent, a deputy schout (municipal official), impounded goats owned by Hendrick the Tailor and two other defendants for damaging private property. The defendants asserted that they were not at fault, as the property owner's fence was in disrepair and any number of animals could have caused the damage in question. The defendants were ordered to pay unless they could prove their claims regarding the state of the fence; they apparently ignored this directive, as a later order required the goats to be sold for lack of payment.²

When the Dutch traveled across the Atlantic and began fur trapping along what is now known as the Hudson River in the early 1600s, they brought their livestock with them. Throughout the relatively short history of the city of New Amsterdam, which existed for a little over forty years until it was taken over by the British in 1664, who renamed it New York, officials were often preoccupied with regulating animals, their owners, and the behavior of both. In 1647, for example, the Dutch developed rules requiring livestock owners to keep animals in fenced areas, threatening to impound the animals and hold the owners liable for any damage caused by the animal in question. Three year later, specific rules were adopted regarding animals in and around Fort Amsterdam, a structure that had fallen into great disrepair, "trodden down by hogs, goats and sheep."³

The Dutch predicament was not unusual; goats caused problems throughout the colonial era. While there is no specific record of animals traveling on the *Mayflower* in 1620 (though goats are thought to have been onboard), visitor Emmanuel Altham noted the presence of at least "six goats and fifty pigs and hogs and divers[e] hens" at the Plymouth colony in 1623. Cattle did not arrive at Plymouth until 1624, escorted from England by Edward Winslow. In 1629, Puritan minister Francis Higginson set forth from England with five ships, including several hundred passengers and cargo, on their way to establish what would become the Massachusetts Bay Colony. In letters he sent back to England, Higginson recorded the presence of a number of animals on the ships, including six goats on the *Talbot*, as well as cattle and "some" goats on the *George*. Goats and a variety of other livestock were also transported to other European colonies, including Samuel Champlain's French settlement in Quebec and the short-lived colony of New Sweden established along the lower Delaware River.⁴

Europeans and their livestock survived and multiplied in the New World. Ralph Hamor, an early resident of Jamestown, noted the presence of "two hundred cattle and as many goats" in 1614, just seven years after the first colonists had arrived there. Hamor was moved to dwell specifically on the fertility of the goats, which bring forth "three and most of them two [kids]," a fecundity he attributed to divine providence. A census taken at Jamestown ten years later enumerated 170 goats and 325 cattle. The enthusiastic publication *A Perfect Description of Virginia*, written in 1649, counted 20,000 cattle, 3,000 sheep, and 5,000 goats, though the numbers were surely exaggerated. Edward Johnson reported the presence of 12,000 cattle and 3,000 sheep at the Massachusetts Bay Colony by 1650 and noted hundreds, if not thousands, more in his descriptions of the surrounding communities. By the mid-seventeenth century, domesticated animals of all kinds were so numerous that their numbers far exceeded that of colonists.[5]

Goats were a relatively common, if not always welcome, sight in colonial America. John Josselyn, an Englishman who visited New England several times during its early years, remarked on their presence in an account of his travels, noting, "Goats were the first small cattle [English settlers] had in the country; he was counted no body, that had not a trip or flock of goats." The practice was fairly common; John Winthrop Jr., governor of the Connecticut Colony, kept a number of animals, including horses, sheep, and several hundred goats on Fishers Island off the coast of Connecticut. Islands were quite useful during this period for livestock keeping because the surrounding water provided a natural barrier that helped contain the animals, as well as providing some protection from predators.

While goats were useful during the early days of acclimation, most early European colonists were glad to be rid of them once cattle began to reproduce and populate the region. Goats were generally considered an animal of the poor and most appropriate for those without other means. In an account of a trip to New Netherland in 1641, Adrian van der Donck described the use of goats by Dutch colonists:

> The inhabitants keep more goats than sheep, which succeed best... Goats also give good milk, which is always necessary, and because they cost little, they are of importance to the new settlers and planters, who possess small means. Such persons keep goats instead of cows. Goats cost little, and are very prolific[.]

Although many colonists were skeptical about goats, there were advantages to keeping the animals. For one, goats are prolific. The gestation period for goats is roughly half that of cattle, and they typically deliver at least two offspring, as Ralph Hamor had enthusiastically noted. Goats also are able to reproduce within their first year of life, much sooner than cattle. In European settlers' first few years living in unfamiliar surroundings, goats provided meat and milk while cattle grew to reproductive age. Goats were also useful for assistance in land clearing, an arduous task, as most land in the New World was not immediately conducive to European-style agriculture. Unlike other livestock, goats are browsers and readily consume leaves and brush in addition to grass, making them an ideal tool for this purpose (as they are often employed to do in contemporary society).[6]

By far the most important use of goats in the early East Coast colonies was for food, but goat was not colonists' meat of choice. Roger Clapp, an early Boston resident, noted in his memoir that, in the early days, when colonists were scrambling to survive, they had to "be contented with mean things . . . it would have been a strange thing to see a piece of roast beef, mutton or veal; though it was not long before there was roast goat." William Wood attempted to put a positive spin on the necessity of consuming goat's meat by asserting that it was "altogether as good [as mutton] *if fancy be set aside*" (emphasis added). Even so, goat meat was featured on colonial tables: a traveler to Jamestown in 1634 found "tables furnished with pork, kid [young goat], chickens, turkeys, young geese . . . besides plenty of milk, cheese butter and corn[.]"

One dish often consumed in the early New England colonies was called samp, a porridge composed of ground corn cooked with water. If milk was available, it was typically added to the corn mixture; it's reasonable to think that a number of men, women, and children may have consumed goat's milk samp at one point or another in the New England colonies during the earliest years of acclimation. Given the presence of goats, it is also possible that some form of fermented goat's milk product or cheese could have been made and consumed, perhaps even exchanged or sold, during the seventeenth century.[7]

❖

Thomas Minor was an English bachelor who migrated to New England in 1632. He eventually became a farmer and acquired a plot of land near Stonington, Connecticut, about ninety miles south of Boston. Minor's farming endeavors are helpful in understanding the nature of agriculture in the New England colonies because he recorded his experiences in a diary, one of the few sources of this period that detail everyday life and practices on a farm. Minor kept a variety of livestock, including cattle, horses, and goats. As was common practice, Minor's animals mostly wandered without supervision during the temperate months, causing Minor and his sons to often spend time looking for "lost cattel" and swine.

Goats first appear in Minor's diary in 1655, when he notes, "Tuesday the 15th [of September] the goats were lost." Over the next several years Minor details the birth of baby goats (kids): in October 1657 he records forty kids, an indication that his herd likely included at least fifteen or more females. Several days later, some of those kids and lambs were killed by wolves. Livestock appears to have been a significant source of income for

Minor, who periodically traveled to nearby livestock markets in Boston, Massachusetts, and New London, Connecticut, to sell cattle and horses. In one instance, he exchanged several goats with one Mr. Perke for a colt. Minor does not discuss milking goats or eating goat meat; his notes are confined to livestock raising rather than dairying or cooking, which at the time were considered women's work.

The trajectory of Minor's goat keeping practices abruptly changes in 1661, however, when he notes in his typical terse prose that "we had all our kids killed Monday." Though the diary continues for several more decades, Minor never mentions goats again. While his silence is not an absolute indication of the absence of goats, by killing all of the young goats Minor at minimum eliminated future breeding stock, more or less forestalling goat keeping going forward. Minor continues to detail his experiences with growing crops and keeping cattle and horses, however, including periodic treks to livestock markets to sell the animals.[8]

Thomas Minor faced a different set of challenges than the starving European emigrants of earlier generations. By the 1670s there were some 50,000 colonial residents in New England and 40,000 in the Chesapeake Bay region. No longer in survival mode, settlers on the Atlantic Coast were beginning to establish a society resembling what they had been accustomed to in Europe. As cattle increased in numbers, value, and importance, goats began to fall out of favor. While goats were useful in the early days of settlement, the population soon began to decline significantly. The reasons were in part pragmatic; goats gave less milk than cows and provided less meat. A reasonably established seventeenth-century colonial farmer like Thomas Minor could make a simple economic calculation: why waste time on goat keeping when cattle provided more of both?[9]

But there was another reason for the decline in popularity of goats over the course of the seventeenth century: the animals' (perceived) destructive habits. Across New England, goat behavior came to merit particular attention by civic authorities. In 1641, for example, the goats belonging to Massachusetts resident William Brown strayed near Mr. Batter's farm. Mr. Batter was forced to release his "great dog" on the goats. The dog killed one of the goats, an offense for which Mr. Batter was ordered to pay. Robert Goodell was fined in 1644 for allowing one of his goats to stray into his neighbor's cornfield. Across jurisdictions authorities adopted regulations to contain the goat problem. For example, officials in Cambridge, Massachusetts, wrestled often with the issue, finally passing an ordinance in 1639 that stated:

> Ordered, for the preservation of apple-trees . . . and for preventing all other damage by them and harm to themselves for skipping over pales, that no goats shall be suffered to go out of the owners' yards without a keeper; but if it appeareth to be willingly, they shall pay unto anyone that will put them to pound, two pence for every goat[.]

Like neighboring Cambridge, Boston was also besieged with problem goats. In 1641, commissioners noted, almost wearily, for the record "the great damage done by goates unto gardens, orchards and cornfields; the great grievance that often ariseth among the Inhabitants by reason of them, the many orders made about them, and yet altogether ineffectual," and threatened a total ban if existing rules were not observed. Similarly, in 1645 officials in New Haven, Connecticut, instituted a fine that was required to be paid by goat owners if goats were found "in any street, way or lott . . . without a keeper."[10]

By the mid-seventeenth century, goats were legislated to the margins of life in much of New England. Boston banned goats entirely in 1642, "never more to be seen again, under [monetary] penalty . . . for every goat seen abroad from this day forward," though provisions for access to cow's milk were made for the poor who relied on the goats for milk. In 1650, Warwick, Rhode Island, made it lawful for citizens to kill goats caught straying onto the town commons. Even farmers in less populated areas outside of towns and villages had little interest in goats: "the people most inclining to husbandry have built many farms remote [in rural Massachusetts]; cattle exceedingly multiplied; goat[s], which were in great esteem at their first coming, are now almost quite banished."[11]

It's worth dwelling briefly on the fact that many colonists' complaints about goats centered specifically around the damage or destruction of fruit trees. Fruit was serious business in the early colonies, especially the cider made (typically by women) from apples, pears, and other fruits. Colonists brought their taste for alcohol from Europe, where water was a dangerous commodity; lakes, rivers, and streams were contaminated by garbage, sewage, animal carcasses, and other assorted debris, which encouraged the spread of diseases like cholera, dysentery, and typhoid. As a result, alcohol was consumed in great quantities by men, women, and even children and was considered healthy and nourishing. In the New World, however, producing alcohol was no simple prospect since barley and other grains necessary for brewing and distilling were not always available. Fruit trees, by contrast, grew essentially independently compared to the intensive labor associated with growing and harvesting wheat or other grains. Cider production required just the pulverizing of fruit and the fermentation of the juice, which occurred spontaneously anyway. When it came to precious fruit trees, goat misbehavior was something most people could agree on.[12]

The general decline in goat keeping across many of the American colonies was in part reflective of prevailing attitudes toward goats back home in England. Historically, goats were fairly common across the British Isles, but studies have shown that the goat population plummeted during the early Middle Ages. Researchers attribute the decline in part to the evolution of organized intensive agriculture, which favored more productive and profitable animals like cattle and sheep. In addition, the rise of the feudal system drove the enclosure of lands once used for common livestock grazing; this may have contributed to the decline of goats, as a common complaint of the period was that hungry goats destroyed hedgerows separating privately held properties. While a few large English estates continued to maintain their own goat herds, by the fifteenth and sixteenth centuries, around the time English efforts to colonize North America began, goat keeping had been relegated largely to the peasant classes. As a result, English colonists were generally predisposed to dislike goats, and they brought those beliefs and associations about the animals with them across the Atlantic.[13]

Goats in Nineteenth-Century Cities

During the nineteenth century, the city of San Francisco had a goat problem. A local newspaper provided the details: "residents... complain bitterly of the ravages committed on their shrubbery by herds of goats, which infest [the area], roaming hither and thither, and devouring whatever comes their way. No fence is a protection against the nimble footed thieves[.]" Goats roaming through the city's streets caused already frayed tempers to boil over. An outraged citizen complained of "[n]umerous flocks of goats and sheep of all ages ... leaping over and even breaking down gates and railings and gratifying their simple tastes." In a more serious incident, a frustrated man shot at a goat running through the streets, almost hitting a bystander; the near victim sued.[14]

San Francisco was not the only American city with a goat problem. Goats ranged freely through the streets of cities up and down the Eastern Seaboard, from New York to Baltimore and Washington DC. In nineteenth-century Baltimore, residents complained bitterly and often about stray goats that "enter[ed] any premises they find open" and destroyed their yards. "Shrubbery or garden vegetables stand but a poor chance if [goats] find their way in. And indeed, they will get where clothing is hung out to dry and will eat them into holes." Goats were also a regular part of life in Philadelphia; according to one report, a nine-year old boy was injured as he attempted to pet a goat. The goat's horns just missed severing the boy's windpipe. In Washington DC, a colorfully described "reckless chamois of the Alps" broke through the front window of a house, intending to eat the plants visible inside. James Maher, who bore the title of United States public gardener, wrote a letter to *The Republic* newspaper complaining about the stray goats destroying chestnut trees he'd so carefully planted along Pennsylvania Avenue.[15]

Since the earliest days of European colonization, domesticated livestock were a primary basis of human sustenance and livelihood in what is now the United States. The American economy developed around agricultural production, and wealth was typically determined by the land and animals that a family owned. But starting in the nineteenth century, everything changed. The Industrial Revolution, already well underway in Europe, began to reshape the United States economy. The earliest textile mill, a cotton spinning factory, opened in Rhode Island in 1790, and New York City soon became the center of sugar refining and garment manufacturing in the United States. Technological innovations such as the steam engine powered the transformation of industrial-scale manufacturing of all kinds. Factories required labor, and populations in cities up and down the Eastern Seaboard skyrocketed as immigrants arrived from all over the world. New York was on the vanguard of the nation's shift toward industrialization; already the nation's largest city, New York grew from a population of 590,000 people in 1850 to over 1.9 million in 1880, an increase of over 200 percent.

The nation's rapidly expanding cities gradually began to encroach on the fields and farms that once defined the American colonial landscape. The transition was neither fast nor efficient. For many years, cattle, pigs, sheep, and goats roamed the streets of metropolitan areas, just as they might have once roamed through villages and rural areas. Not surprisingly, loose livestock became an increasing burden on life in densely packed urban landscapes, and municipal officials across the country grappled with animal-related issues for many decades. Urban growth and expansion during the nineteenth century forced a negotiation over the present and future character of urban spaces.[16]

The story of urbanization, industrialization, and problem goats evolved differently on the West Coast, though the result was the same. There, the goat story began with the Gold Rush. In January 1848, carpenter James Marshall discovered gold while constructing a sawmill at John Sutter's property in the Sacramento Valley—the Gold Rush was on. Thousands from across the United States and as far away as Peru, China, and Australia left their homes, families, and livelihoods to traveled to San Francisco, the nearest port to gold country, in hopes of striking it rich—a local newspaper aptly termed the phenomenon the "Golden Immigration." The city's population exploded, growing to over 15,000 by late 1849 and then to 36,151 by 1852, an increase of over 140 percent. Quite suddenly, San Francisco became the West Coast's first big boomtown."[17]

While gold seekers traveled to California seeking precious metals, others found opportunity in supplying the rapidly accumulating masses. One particularly lucrative business venture was the livestock trade, as the sudden, heavy demand for food and transportation sent the value of cattle and other domesticated animals soaring. So lucrative was the trade in livestock that a parallel "livestock rush" emerged, as hundreds of thousands of cattle, sheep, and horses were transported from all over the continent by a variety of means to Northern California. And so, along with the thousands of people making their way to the region, came many thousands of cattle, sheep, and goats. By 1855, the newly formed California legislature reported the presence of over 500,000 cattle and nearly 200,000 sheep in the state.

Though goats were not high-volume, high-profit livestock like cattle and sheep, they also found their way to California. Since sheep and goats were typically herded together in the Southwest, many goats traveled among the hundreds of thousands of sheep driven to the region. Goats were also transported to the region via ship, and local newspapers carried news of arrivals along with details of incoming cargo. The *Daily Alta California* announced the arrival of two hundred sheep and goats along with four hogs on the *John Ender* in June of 1851. Others advertised their goods in newspapers: "To Farmers: For Sale, 200 goats, the finest race ever brought to San Francisco, including a large number of milk goats and kids... apply on board Schooner Honolulu, foot of Market St.[.]"[18]

❖

Across the country, fast-growing cities struggled to deal with the wide variety of livestock in their midst including cattle, pigs, and, of course, problem goats. In San Francisco, residents petitioned city officials as early as 1855, pleading with them to do something about the destructive stray goats and other animals. An ordinance was eventually adopted "prohibiting the running at large of goats and hogs within the city." These types of enactments were a common early tactic in fighting urban goat problems. Decades earlier, in 1790, the city of Philadelphia adopted a similar ordinance, which stated that for "any goat found going at large in the public street or alley in this city, the owner of every such goat shall pay a fine of a half dollar." Likewise, the New York City Common Council passed an 1804 ordinance declaring that "if the owner or owners of any goat or goats shall suffer or permit any such goats to run at large in any streets or alley in this city . . . such goat or goats shall be forfeited," with the owner required to pay ten dollars to recover the animal."[19]

Although the passage of laws addressing goat-related issues showed that aggrieved citizens had captured the attention of municipal officials, laws generally did little to solve the goat nuisance problem because there was generally little to no effort directed toward enforcement. "There is an ordinance of the city which provides that goats shall not be allowed to go at large," *The Baltimore Sun* newspaper complained in 1846, voicing the sentiments of frustrated city dwellers across the country, "Why is the ordinance not enforced? There are portions of the City where these animals are a positive nuisance."[20]

Enraged citizens often took matters into their own hands. In one instance in San Francisco, aggrieved homeowner J. P. Newmark paid a young man one dollar to get rid of a goat that had been troubling him. Newmark was subsequently sued by the goat owner, and a judge ordered him to pay ten dollars for the loss of the goat in question. Commenting on the Newmark case, a newspaper spoke for many in the city when it observed that "the courts seem to act on the principle that citizens have no rights which the goats are bound to respect." Frustrated residents of New York's Yorkville neighborhood, which some irreverent citizens called "Goatville," went on the initiative against the roaming goats in their neighborhood, forming an Anti-Goat Protective Association to address their neighborhood's goat ills in the absence of any effort from the city. Others took more drastic measures, setting out strychnine in the streets in the hopes of poisoning the animals. The latter approach could have dire consequences, however; in one unfortunate case in New York City, a plate of arsenic combined with sugar, thought to have been left out for goats, was consumed by a child who later died.[21]

The proliferation of animal problems of all kinds forced officials to devote more concentrated attention to animal-nuisance issues. One of the more promising means of addressing issues caused by goats and other animals was the creation of institutions specifically directed toward animal control and containment. But the solution was not that simple, as

Corner of Mott and Hester Streets in New York City, 1902. Photo by Robert Fellows Wood (44.341.30). Collection of the Museum of the City of New York.

rampant corruption often hampered such efforts. In San Francisco, the most notorious of a succession of San Francisco poundmasters was Jake Lindo, an acolyte of "Blind Boss" Buckley, the Democratic Party boss who controlled San Francisco city politics during the 1880s. Lindo's army of toughs grabbed pets directly out of children's hands in an effort to pocket the fees required for recovering animals from the pound. In one instance a man was assaulted by two pound deputies as he tried to prevent them from snatching his two goats. A corruption investigation eventually revealed that Lindo had been siphoning off carcasses of dead animals to friends, who profited handsomely from the sale of the meat and hides. Ultimately the Lindo debacle reminded frustrated citizens that they could not rely on city officials to solve their animal problems.[22]

While animal shelters and other types of containment facilities helped to mitigate urban goat issues, the more challenging problem proved to be capturing the animals. Newspapers delighted in chronicling of the adventures of police officers and other officials in their attempts to capture goats. In a Brooklyn neighborhood, police officers responding to a burglary call descended the steps of the residence, only to be immediately butted by a goat "at the fourth button of his official waistcoat." After a brief battle with the goat, the policemen left the residence, seeing no burglary; according to reports, the woman who had summoned the police laughed heartily at the officers. In Philadelphia, a brave fireman managed to rescue a woman from being attacked by a goat; the fireman "had once attended a college football game and . . . knew how to tackle anything from a goat to a locomotive." The fireman was only partially successful, however, as the goat butted him in the midst of the fracas and a second fireman had to intervene in order to subdue the animal. To be sure, the prospect of rounding up stray, hungry goats was no simple task; one account of a New York police officer's encounter with a group of goats, in which the goats are referred to as "scriptural types of wickedness," detailed that while the officer successfully rounded up twenty-eight goats, he "received forty six wounds in the day's engagement." To the great frustration of citizens and municipal officials, there were usually more goats just around the corner.[23]

Two Goat Hills

Telegraph Hill is situated in the northeastern corner of the San Francisco peninsula. One of many large hills in a very hilly city, Telegraph Hill has been called a variety of names, including Goat Hill and Prospect Hill; the hill acquired its present name when a semaphore, a signaling device used to monitor ship traffic before the invention of the electric telegraph, was constructed on its peak. Once the default home of Gold Rush hopefuls, Telegraph Hill evolved into a poor and working-class immigrant neighborhood because of its proximity to the waterfront at its base, as well as its precipitous slopes and absence of roads—in other words, its inhabitability. The landscape itself was considered rough and expendable; throughout the early history of San Francisco the cliffs of Telegraph Hill were chipped away, and the dirt and rock were used as ballast on outgoing ships.

Telegraph Hill first gained its reputation as "the paradise of San Francisco goats" during the Gold Rush years. And it really was a goat paradise; the steep, rocky hill with its grassy plateau was cumbersome for people and larger animals, but easy for nimble goats to navigate. In 1857 a plucky reporter scaled the Hill, noting its views and the evident poverty he observed along the way. The "tenements you have so often gazed at from a distance [are] not at all improved by the proximity of your scrutiny," he wrote. "You meet herds of dogs and goats . . . Each street exceeds its predecessor in uncleanliness." Because the upper reaches of the hill were sparsely inhabited for many decades, area goat keepers allowed their goats to graze freely on the hill, using it as an informal commons. An author who grew up in the area in the 1850s remarked in a memoir that the top of the hill had served

Telegraph Hill, 1885. Photo by Martin Behrman, copyist; original photographer unknown. San Francisco History Center, San Francisco Public Library.

more or less as a playground for "[the] flocks of goats that browsed there all year round, and the herds of boys that gave them chase."[24]

Numerous nineteenth-century cities had neighborhoods known for their goat population. Among the East Coast neighborhoods known to be centers of goat nuisances were Kensington in Philadelphia, Locust Point in Baltimore, and the colorfully named Swampoodle in Washington, DC. Among the most notorious of these neighborhoods was New York City's Goat Hill, also known as Dutch Hill, an area situated along the East River, extending northward from around Thirty-Nineth Street where the United Nations complex sits currently. The area was once a tranquil inlet along the river during the period when the Indigenous Lenape people lived on the island. Dutch Colonists took control of the territory during the early seventeenth century and subsequently granted title to the area to two Englishmen during the seventeenth century who developed a tobacco farm. Meanwhile New Amsterdam residents used the upper reaches of Manhattan Island for grazing their livestock, including goats.

The earliest industrial development in New York City was centered at the south end of Manhattan Island near already established ports. The area to the north around Dutch Hill was slower to develop in part because it was hilly and rocky. A nineteenth-century geological survey performed before the area was leveled and developed describes prominent cliffs extending "some sixty to eighty feet above tidewater." The high, rocky riverbanks were said to have once made a convenient enclosed harbor for small vessels and later became a playground for goats. Like Telegraph Hill, the Dutch Hill landscape was repeatedly blasted with dynamite and incrementally dismantled; Dutch Hill granite was used in building projects all over the city, including in the early construction of Bellevue Hospital, located just to the south.

View from the school house in Forty-Second Street between Second and Third Avenue, looking north (Dutch Hill), 1868. Note the goats, pigs, and chickens in the foreground. Unknown engraver. PR 020, Geographic Images Collection. New York Historical Society 74193.

Dutch Hill became home to a settlement of poor immigrants that was by contemporary accounts the largest in the city. Residents, among them Irish and German laborers, crafted what were called shanties from scavenged lumber, rags, and other foraged materials; reports note at least 250 or more makeshift dwellings in the area. Impoverished residents kept goats, chickens, and other animals to support themselves, and those same goats often roamed freely through the area and beyond. "Each house has a retinue of large dogs and goats and pigs," according to one description, "The goat seems always an unhappy animal unless he can somehow find himself on an inclined plane, so here he satisfies his disposition by standing in a precarious manner on a sloping cart laid by, or even on the roof of the cabin, from which he looks out complacently on the landscape." Complaints about goats in the area were a constant over the decades. One group of concerned citizens claimed constant harassment from "numerous hordes" of goats that had "assert[ed] their right to family linen and kitchen utensils as luxuries on which all respectable goats must subsist."[25]

The ongoing struggle to contain stray roaming livestock within American cities eventually began to evolve into formal and informal campaigns to contain and control populations of the poor and immigrants who kept the animals. The goat "problem," such as it was, became less about stray animals and more about their owners. Goats evolved into potent symbols of intractable urban decay because they did not have any inherent value and roamed essentially unabated. Communities of poor immigrants became a convenient target for anti-goat sentiment. During the 1880s, San Francisco's Telegraph Hill neighborhood was described as "reflect[ive of] the worst traditions of Old World poverty," a place where "squalid children [sit] on the steep flights of steps or on the cliff-sides, fraternizing with the denizen goats and pigs." One particularly scathing assessment compared the residents of Telegraph Hill to prehistoric cliff dwellers. Telegraph Hill, with its large expanses of grass and still lacking roads, had become known for its population of Italians, many of whom kept goats for milk and meat. During the late nineteenth century, hundreds of thousands of Italians, many from the southern part of the country, migrated to the United States to escape political turmoil and poverty. While many settled in cities up and down the Eastern Seaboard, a significant number also made their way west to opportunities in California. Many in the Telegraph Hill and adjacent Russian Hill areas kept goats; for other citizens of San Francisco, goats became the default symbol of uncivilized backwardness. The phrase "Telegraph Hill goats" became a metaphor for a variety of negative aspects of civic life.[26]

While urban animals were kept by poor and immigrant populations of a variety of races and nationalities, on the East Coast the popular press emphasized an association between goats and the Irish. Irish immigrants began arriving in the United States in large numbers during the Potato Famine of the mid-nineteenth century. By the 1860s nearly two million people of Irish descent had migrated to the United States: many made their living as laborers in cities across the continent. New York City's most notoriously poor neighborhoods of the nineteenth century, including Dutch Hill, Five Points, and others, were densely packed enclaves of mostly Irish immigrants. Even Telegraph Hill in San Francisco began as a mostly Irish neighborhood until Italians arrived in later waves of immigration and settled in the area.

Popular culture of the mid-nineteenth century fixated on the Irish immigrant as the source of urban goat-related ills, and the Irishman and his goat became a general metaphor for a wide range of social problems. Stories about renegade goats were invariably reported to have Irish owners (whether or not the owners were specifically known to be Irish), indicated by deliberately crafted Irish names like "Mrs. Finn" or "Mr. Mackenzie" and other stereotypes of the time, including drunkenness and uncivilized behavior. Satirical publications of the day such as *Puck* and *Judge* ran cartoons showing drunken characters with signifiers of their Irish heritage, which typically included a goat or a shanty lurking in the background.

After decades of wrangling over the presence of goats and other animals in their midst, urban reformers began to envision "civilizing" the immigrant poor as a means of eradicating their nuisance animals. The momentum for solving municipal goat problems evolved from nuisance laws and police roundups toward social work. In San Francisco, a group of reformers focused their efforts on Telegraph Hill and its Italian population. The Telegraph Hill Neighborhood Center was started by Elizabeth Ashe and Alice Griffith, two well-off society women who began working with the poor on the hill in the 1890s, in the tradition of the settlement house movement of the period. "Little Italy is to have the ministrations of more good women, and heaven knows there is a need for it," wrote one reporter, welcoming the potential "social and hygienic improvements" promised by the formal establishment of the organization in 1904. The Telegraph Hill Neighborhood Center focused its attention on the younger generations, offering a variety of programs, including sewing classes, music lessons, and physical fitness classes. Improvement of the people, it was hoped, would have the added effect of eliminating other nuisances, including problem goats.[27]

Among New York's most prominent reformers was Charles Loring Brace, a minister and philanthropist who founded the Children's Aid Society, an organization devoted to removing children from the squalor of urban life, in 1853. Brace took special interest in the Dutch Hill shanties. Among his efforts, Brace enlisted the help of residents of the wealthy Murray Hill neighborhood, adjacent to Dutch Hill, to fund the East River Industrial School, where poor children from the city's shantytowns could be sent to learn skills and trades. By Brace's own account, the children under his schools' care were transformed; he counted as one of his particular successes a "wretched looking man who lived with his pigs and goats."

Shanty Town, Fulton Street in lower Manhattan, 1896. Photo by Jacob A. Riis (90.13.2.137). Collection of the Museum of the City of New York.

The man had two daughters who attended Brace's school; daughter Nellie eventually married and moved to Europe with her husband, "all the old stamp of Dutch Hill quite gone." Nellie had been saved from ever having to live with goats again.[28]

❖

As the twentieth century dawned, the tide began to turn on nuisance goats in cities. A retrospective mused on the relative lack of goats in the upper reaches of New York City:

> The New York goat was the rear guard relic of retiring rusticity. So long as he remained at peace on the cliffs and crags of the ragged, rocky outline of upper New York, the district in which he had his habitat was not accepted as ready for urban settlement . . .

[Now] Upper Fifth Avenue has been redeemed from the goat and dedicated to the glory of civilized man.

Another report noted a similar absence of goats in Philadelphia: "there is a surprising decline in the number of goats within city limits," the author remarked, "In old Philadelphia these horned quadrupeds roamed at will over nearly every vacant lot... they are seen but rarely now[.]" The contrast between the portrayal of these newly civilized urban centers, bastions of progress and civility, and the "retiring rusticity" of the immigrants and their goats that once populated the streets of those same cities could not have been drawn more clearly. The vision of a civilized landscape that separated urban activity and agriculture was beginning to gain the upper hand.[29]

The evolution of the urban planning concept of zoning city landscapes would eventually prove to be the regulatory mechanism by which many cities finally excluded urban livestock of all types. Zoning typically defines specific permitted uses for different areas of a city, most typically industrial or residential, defining cities into areas of separate planned uses. Zoning laws effectively codified the patchwork of nuisance laws that had proliferated throughout the nineteenth century. New York City passed the nation's first zoning laws in 1916 to regulate the building of skyscrapers, and within a few years the trend spread to cities across the nation.[30]

But not all cities celebrated the end of goats. In the aftermath of the San Francisco earthquake in 1906, as many as 250,000 people, rich and poor, lost their homes and livelihoods, and many wandered the streets for months. The city's animal populations were affected as well; stray pets roamed everywhere, among them cats, dogs, and rabbits, some of which were taken in by those sheltering in refugee camps. In the aftermath of the earthquake, some stray goats that escaped the fires were found roaming in the Potrero Hill neighborhood, which had avoided destruction.

The difficulties of regulating loose livestock resurfaced in post-earthquake San Francisco. Mayor John Rolph's office received numerous complaints about livestock staked out to graze in open lots and public thoroughfares. "We . . . have been tormented with persons pasturing horses, goats and cows along our streets and vacant lots and destroying flowers and shrubbery and also committing nuisance on our sidewalks," wrote one concerned citizen. The mayor's office and the pound took a tolerant stance against such grazing practices. "[W]e have had numerous complaints from various districts that the pound is too severe in taking up stock that is staked out. Poor people have complained that this is a very great hardship on them, and the proper solution would appear to be rather difficult." Civic authorities in San Francisco attempted to walk a fine line, not always successfully, between neighborhood concerns and the needs of the city's poor population.[31]

After years of wrangling over problem goats, the San Francisco Board of Supervisors passed a law known as the Goat Ordinance in 1920. The law explicitly allowed no more than two goats to be kept per family within the city without a permit; this remains the law regarding goat keeping within the San Francisco city limits. But despite the supervisors' attempts to eliminate problem goats, the problems persisted. In the Potrero Hill neighborhood goats continued to wreak havoc, consuming bread dropped off on front porches as well as "eating the tops off of milk bottles and drinking the milk." In 1922, city police promised to start patrolling the neighborhood in the early hours in order to prevent the goats from causing trouble. Several women in the Telegraph Hill neighborhood kept goats and herded the animals to nearby vacant lots to graze well into the 1920s. One of these women was Millanella Cosenza, an Italian immigrant who was known to stroll through the streets of Telegraph Hill with her goats selling goat's milk to order, straight from the animal.[32]

❖

Goats and other domesticated livestock have populated the Western Hemisphere for centuries. While goats were useful in the early days of European colonization when colonists needed them for food, they were quickly cast aside in favor of cattle. Over time, colonial villages grew into urban centers of commerce and production. Across the country, fast-growing cities struggled to keep up with extraordinary population growth and the sweeping social and economic pressures that accompanied it. Goats and other livestock were kept by the large populations of the poor and immigrants struggling to gain a foothold in the fast-industrializing economy. The story of goats in nineteenth-century American cities is the story of the poor and working classes who kept goats as a means of survival in cities, city authorities and their ongoing struggles to contain goats, as well as the tenacity of the animals themselves, which roamed cities in search of food and were singularly adept at eluding measures to control them. By the twenty-first century a new generation of city leaders would find themselves reimagining restrictive laws as a new era of urban agriculture, and urban goat keeping, began to take hold. Goats would eventually return to cities, and many residents would actually welcome them.

CHAPTER TWO
Navajo Goats
Herding and Administrative Control

> It is said the white men make use of everything about the pig but its squeal—[all] of the chicken but its crow. We make use of everything about the goat but its smell.
>
> Mrs. Kee McCabe, Navajo spokesperson, 1931

If you were to create a map highlighting areas of the United States with the largest goat populations, a few regions would stand out. The state of Texas currently has the largest overall number of goats within its borders; the majority of goats in that state are raised for meat. While Wisconsin is well-known for its dairy cows, that state is also home to the largest concentration of dairy goats in the nation. Another large population of dairy goats can be found in Northern California. But there's another region that stands out for the size of its goat population—the desert Southwest. One of the largest populations of goats in the United States currently resides within the boundaries of the Navajo Nation. The substantial number of goats within the boundaries of the Navajo Nation reflects a centuries-long shepherding tradition, a way of life that has been repeatedly challenged by Western culture and values.

First Goats

The Diné creation story, as communicated through the sacred Blessingway ceremony, tells of the emergence and evolution of the People, livestock, and the land around them.* After the People passed through a series of worlds, the outlines of contemporary society emerged. The powerful deity Changing Woman, sometimes called the mother of the Diné, prepared

---

* The Navajo often refer to themselves as Diné, or "the people," though the terms Navajo and Diné are often used interchangeably.

the earth by providing sheep and goats, along with the plants and grasses that cover the landscape with which they could sustain themselves. "The creators gave the People livestock when they were ready to weave and utilize the animals in all of the ways they do today," said Aretta Begay, former executive director of Navajo Lifeway Inc. From there the Navajo grew and prospered.[1]

The Navajo often say, "sheep is life." The expression only scratches the surface of the complex and interdependent relationship between the People, their livestock, and the landscape upon which both reside. Sheep and goats are central to Navajo identity; the relationship between people and animals transcends concepts like property and monetary value through which Western culture typically views livestock. Livestock were given to the Navajo by the creators, and the connection is a deeply spiritual bond manifested in day-to-day mutual coexistence and interdependence. The relationship is an expression of the ideal state of balance and harmoniousness between the people, land, and the natural world, which the Navajo call *hózhó*.

The Western narrative of Southwest livestock begins in 1598, when Spanish nobleman Juan de Oñate, the son of a wealthy silver mine owner who founded the city of Guadalajara, Mexico, was chosen to lead a traveling contingent northward from New Spain into the interior of the North American continent. Rumors of wealthy cities rich in gold and other treasures to the north of New Spain called the Seven Cities of Cibola carried special significance for some Spaniards, who believed the cities were founded by Catholic bishops who fled the takeover of the Iberian Peninsula by the Moors in the 700s. Several prior expeditions, most notably that of Francisco Coronado during the 1540s, had already ventured northward in search of the cities without success. An estimated seven thousand domesticated animals accompanied Oñate, including cattle and sheep, horses, donkeys, and more than one thousand goats, which formed an extended caravan said to have been more than two miles long. Oñate and his group eventually established the Spanish colony of New Mexico, the earliest permanent settlement of Spaniards and their livestock in the region.

During the sixteenth century, the region now known as the Southwestern United States was populated by a variety of Indigenous cultures; some experts estimate the Native population in the region to have been around eighty thousand people. Among them were the Acoma, Zuni, Taos, and others that Spanish colonists called Pueblos, who lived in highly developed communities along the Rio Grande. The Spanish referred to other, more nomadic Native groups throughout the region as Apaches, and others loosely based in northwestern New Mexico the "Apaches de Nabaju," using a Pueblo term describing the farms in that region. These are the people who became known as the Navajo.[2]

The influx of Spanish colonists and their livestock transformed the region. The Spanish instituted a system of forced labor, known as *encomienda*, which conscripted Pueblos to work for the Spanish and required them to pay tribute for the privilege of doing so. Pueblos labored in Spanish fields, performed household duties, and herded livestock for colonists and Franciscan missionaries, who hoped to convert the Native population to Christianity. Archaeological studies have demonstrated the ways Native diets in the region shifted under colonial control from wild game to meat from domesticated animals such as sheep, goats, and cattle, especially where Native villages were in proximity to Spanish settlements and missions. Pueblo weavers replaced traditional cotton with wool from Spanish sheep in their weaving practices. And as the colonial livestock population grew, the non-native species began to strain the region's environmental resources. Cattle, sheep, goats, and horses occupied land and consumed forage that otherwise would have been the domain of native species, which were pushed outside of their accustomed ranges. Letters traded among Spanish officials during the early colonization period express concern about potential overgrazing of the already sparsely vegetated desert landscape, as well as the lack of water needed to support their ever-expanding herds.[3]

Eventually the Indigenous peoples of the Southwest began to keep a variety of domesticated animals themselves. A 1758 Spanish census recorded over 8,000 Pueblo people living in the area surrounding present-day Santa Fe, New Mexico. The livestock population was substantial, including an estimated 16,000 cattle, 7,000 horses, and 112,000 sheep and goats, the majority of which were attributed to Pueblo households. Because the Navajo lived outside of Spanish control, we know comparatively little about their early herding practices. A brief glimpse of Navajo pastoralism was provided by Spanish soldiers during this period, who reported to their regional governor that they had observed the Navajo with small flocks ranging in size from fifty to two hundred sheep and goats, as well as evidence of woven materials.

Over the course of the next century, domesticated livestock continued to support and sustain the Navajo people. But soon a new colonizing power would appear on the scene—the United States.[4]

❖

By the nineteenth century, the political winds began to shift in the Southwest. After Mexico won independence from Spain in 1821, the region fell under the control of the Mexican government. After the Mexican-American War ended in 1848, Mexico ceded large portions of its territory, including the present-day states of California, Arizona, and New Mexico, to the United States. The change of territorial control had significant implications for the Navajo and other Native groups in the region. From the American perspective, Native populations were "savages" whose presence hampered the progress of white civilization. Just a few years earlier, in 1830, President Andrew Jackson had signed the Indian Removal Act, which authorized the removal of Native tribes throughout the United States to land west of the Mississippi River that the government designated Indian Territory. By the time the United States took control of the Southwest in 1848, the United States government had already displaced over one hundred thousand Indigenous people, who were forcibly relocated to new and unfamiliar homes.

After the United States gained control of the Southwest, American settlers swarmed into the region. The United States government established New Mexico Territory in 1850, which covered most of what are now the states of Arizona and New Mexico. Over the next several decades conflict raged between the Navajo, a growing population of American ranchers, and the US Army. By the 1860s the US Army gained the upper hand. The army forced the Navajo to relocate 350 miles east of the Bosque Redondo Reservation at Fort Sumner, New Mexico, a place the Navajo refer to as Hwéeldi (place of suffering). The forced migration, what the Navajo call the Long Walk, represented another chapter of US Indian removal policy. Despite the efforts of the US Army, some Navajo evaded capture. Navajo elder Ron Garnanez related a story told to him by his great-grandmother: "When the soldiers entered Canyon de Chelly to round up [my great-grandmother's family], her father took her and her two sisters to the rim of a canyon and told them to run and don't come back . . . they took a few of the sheep with them, she said somewhere around fifteen . . . she said the animals saved them from the soldiers because when [the soldiers] were getting closer the sheep would start to get nervous and move away and they followed the sheep wherever they went. She told us to always take care of the sheep."[5]

The United States government confined about 8,000 Navajo men, women, and children at Hwéeldi for about five years. The rules of confinement were strict: the Navajo were prevented from keeping livestock or foraging for accustomed foods such as yucca fruit, wild onion, and potatoes. Instead, the army provided them with food rations consisting of substances previously unknown to them, including white flour and coffee. Many Navajo grew ill from consuming unfamiliar foods, and hundreds died from starvation as well as smallpox and dysentery epidemics, which raged through the reservation during the years of confinement. Deprived of their land and livestock, the Navajo suffered physically and spiritually.

Although army officials pressured the Navajo to take up residence in Indian Territory, the Navajo eventually negotiated a treaty that enabled them to return to their homeland in 1868. In article 12 of the treaty, the US government agreed to supply the Navajo with fifteen thousand sheep and goats, five hundred cattle, and one million pounds of corn upon their return. The allocation of livestock, along with those kept by Navajo who had evaded capture, formed the nucleus of the Navajo's future.

Many Goats

The boundaries of the ancestral homeland of the Diné are traditionally defined by four sacred mountains: Blanca Peak to the east, Mount Taylor to the south, the San Francisco Peaks to the west, and Hesperus Mountain to the north. The area between these mountains defines the territory granted by the creators, known as Dinétah, and encompasses parts of the present-day states of New Mexico, Colorado, Arizona, and Utah. When the Navajo returned to the region in 1868, the boundaries of their ancestral home had been rewritten, circumscribed into an area defined by the United States government as a reservation covering just a fraction of the Navajo's traditional territorial boundaries.

Navajo life and culture continued under the administration of a United States government agency known as the Bureau of Indian Affairs (BIA). BIA officials subsequently instituted a variety of programs designed to shape what it determined to be the most prudent course forward for Navajo livelihood, devoting particular attention to Navajo agricultural practices. The BIA perceived Navajo livestock as insufficient on a number of levels: the animals were of undetermined breeds and were not managed to maximize production of profit-generating commodities, particularly wool. As a result, the BIA initiated programs designed to improve Navajo livestock and align them with the standards and practices of Western society.

Toward that end, the BIA introduced a variety of breeds of purebred livestock to the Navajo reservation. Merino sheep arrived as early as the 1870s, and over the decades the BIA added additional sheep breeds, including Rambouillet and Cotswold. The purebreds were crossed with reservation sheep in order to develop offspring that would produce a larger quantity of desirable fleece. From the BIA perspective, the crossbreeding programs showed promise; an early twentieth century report noted that "the result of the first cross between native sheep and high-grade rams is the production of an animal 20 to 25 percent greater in weight and yielding 50 to 60 percent more wool of a considerably better grade." Improvement efforts continued well into the twentieth century; in 1935, the BIA established the Southwestern Range Sheep and Breeding Laboratory at Fort Wingate, New Mexico, a facility dedicated to developing a desert-adapted sheep breed with wool and meat traits optimized for commercial sale. The lab closed in 1966.[6]

BIA officials also focused some attention on goat improvement. Angora goats were first formally imported into the United States from Turkey in 1849, and by 1900 there were an estimated 30,000 Angoras in New Mexico and 10,000 in Arizona. Angora fiber, known as mohair, became particularly valuable in the later nineteenth century, when the rapidly expanding railroad industry began to use mohair fabric for train seats. BIA officials encouraged Navajo herders to shift to herding Angoras as an alternative to the prevalent "scrub goats" of Spanish ancestry. Several BIA subdivisions of the reservation acquired purebred Angoras; the Northern Navajo Agency purchased fifty-one purebred Angora bucks "to do to the goat herd precisely what has been done to the herds of sheep . . . and through the goats provide the Navajo with another source of income." BIA efforts led to a gradual transformation from the existing Navajo goat population to the more commercially valuable Angora breed. In 1922, the Northern Navajo Agency purchased ten Toggenburg bucks, a prominent breed of dairy goat, and several years later suggested that they could be crossed with the Navajos' growing population of Angoras to develop a hybrid goat useful for both subsistence (meat and milk) and marketability (the wool or fiber). Purebred dairy goats do not appear to have been introduced to the reservation in a broad systematic manner like Angoras, however.

In the decades after the Diné returned to their home territory, both the human and livestock populations rebounded. By 1881, the livestock population had grown to 1,000,000 sheep and goats. While severe drought conditions during the 1880s and '90s drove down the large population numbers significantly, the herds rebounded, and by 1915 the estimated combined population of sheep and goats increased again to an even more remarkable 1.8 million animals.[7]

❖

While the Navajo kept sheep for both wool and meat, goats were also a significant component of traditional Navajo foodways. Navajo stories and oral histories detail the uses of goats for both milk and meat. While often consumed on its own, goat's milk was also combined with other available foods, such as cornmeal or wheat flour. Numerous oral histories mention a favorite meal consisting of cornmeal mixed with goat's milk, sometimes cooked into cakes. Kneel down bread is a popular dish made during harvest season from freshly harvested and ground corn wrapped in corn husks and baked in the ground. The result resembles a Mexican tamale; according to some accounts kneel down bread was served warm in boiled goat's milk.[8]

The Navajo also regularly made cheese from goat's milk. In Western traditions, cheesemakers use several types of substances to coagulate milk solids in order to make cheese. In some countries, including Spain and Portugal, plant-based materials from certain varieties of thistles are used for this purpose. Others employ rennet, a substance derived from the stomach of a very young ruminant mammal. The Navajo used both methods in their cheesemaking practices. Navajo elder Nanabah Begay recalled in an oral history that her grandfather would boil goat's milk and make cheese, which he would hang to drain the whey. In his autobiography *Navajo Blessingway Singer*, Frank Mitchell describes another cheesemaking process. Family members would "pick the weeds . . . with the round leaves, purple flowers and yellow berries that are about the size of marbles [likely the silverleaf nightshade plant, which is plentiful in the region] . . . we would put [the plants] out in the sun to dry and keep them through the winter. Any time during the year we could put those dried berries right into the milk. Their juice makes cheese form on the milk." The curds were formed by hand into balls, which were set outside to dry in the sun. Other accounts describe producing cheese using rennet derived from the stomach of a baby lamb or goat.[9]

By the turn of the twentieth century, demand for goat's milk was growing across the greater United States (see chapters 3 and 4). While the emerging community of American goat enthusiasts favored European dairy goat breeds for milk production, the cost to purchase one or more was out of reach for the average person. A purebred Toggenburg doe, the preferred breed of the period, could cost $200 or more; that's nearly $3,000 in contemporary dollars. An animal with some European breeding, though not purebred, cost about $50–75 ($1100). These prices were enormously high if you consider that the average US income in 1915 was just $687 per year. A relative shortage of European dairy goat stock, coupled with a strong demand for goat's milk, caused goat breeders to seek alternative populations of goats to supplement their herds—and some looked toward the Navajo reservation.[10]

Through the eyes of some Anglo goat fanciers, Navajo goats were a desirable commodity, particularly because they were already used for dairying. Though these were not the preferred European breeds, they rationalized that animals grazing on reservation lands represented a potential resource for building their dairy goat herds, "as these goats furnish the sole milk supply for the Indians a sort of negative selection has been practiced by killing off for meat and hides the poorest milkers and leaving the best milkers in the flock for future breeders" (though selecting the best milkers was their own practice as well). Others disapproved of Navajo goats, believing there had been "no rational system of breeding" of the animals on the reservation. These skeptics looked instead to large populations of wild goats living on the Channel Islands off the coast of Southern California as well as on Guadalupe Island to the south, off the coast of Baja California, Mexico, to stock their herds, but these populations also had their issues; some found that the "island goats" were too wild, with poor udders and short lactation cycles.

Despite the professed interest, outsiders did not view the Navajo goats as truly desirable in and of themselves; the animals merely represented a convenient blank slate for breeding purposes. Once a Navajo goat was bred to a purebred animal, the resulting offspring would then carry 50 percent purebred genetics; as the thinking went, the offspring would be a goat with better European dairy goat genetics, and it would bring a higher price if sold. If a breeder had the time and the means, they might continue the exercise by breeding that 50 percent European goat with another purebred and, over several cycles, reach a point where the offspring approached 100 percent purebred genetics. While they were not exactly the genuine article, as purebred goat owners would be quick to point out, they approached to the highly sought after productivity of purebred dairy goats.

From the perspective of Southwest livestock traders, Navajo goats were another asset in a livestock-rich region. During the 1920s, the Navajo Livestock & Trading Company (not affiliated with the tribe), offered a train carload of what it called Navajo goats to interested California goat dairy farmers for $5 per head. Some advertised Navajo goats for sale in dairy goat magazines and in newspapers. "170 head of young Navajo goats for sale," read one ad in a New Mexico newspaper, "priced right for quick sale." In one instance, a Utah entrepreneur purchased 125 goats directly

Native Navajo Milk Goats

These goats for generations have been used by the Indians for milk, meat and hides, and many milk two quarts and over. Lactation period, 6 to 8 months. They come in a wide variety of colors—white, black, brown, tan, etc., and can be selected for Saanen, Nubian and Toggenburg crosses, so that the progeny will show its breeding to a marked degree as well as generous milk yields.

For further information, prices, etc., address **C. B. ALLAIRE, San Antonio, New Mexico.**

Ads like this for Navajo goats were featured in goat publications during the early twentieth century. From *The Goat World*, January 1920.

from the Navajo reservation and sold them to Greek miners in that state, a population hungry for a familiar food from home. Immigrants from Greece, Italy, and other countries who traveled to the American West for jobs in the mining and other industries often kept goats to help feed themselves and their families. In some cases, they started businesses selling goat's milk and cheese to fellow immigrants. (See chapter 4). No doubt livestock traders played fast and loose with the term "Navajo goat," as few buyers would have been able to distinguish between a true Navajo goat from the reservation and a similar Spanish goat raised by Anglo ranchers or Mexicans outside of the reservation boundaries. At least one (purported) Navajo goat made it as far as the Atlantic Coast: in 1921 *The Boston Globe* ran a story about John Paine of Franklin, Massachusetts, who counted a Navajo goat among his herd. Although not as valuable as his purebred Saanens, and likely kept as a curiosity, Paine declared that Rosie was "the best old goat in the country."[11]

Too Many Goats

Drought conditions have affected the Southwest for centuries. Through analysis of tree ring formation patterns, scientists have been able to determine that severe and prolonged dry periods have occurred repeatedly in the region since at least the 800 CE. During the late nineteenth century environmental conditions became increasingly serious after successive periods of drought during the 1880s and '90s. The US Agricultural Census observed in 1900 that "the flocks of the Navaho have suffered severely in the last ten years on account of the steadily decreasing rainfall and resulting scarcity of vegetation. In some seasons the almost total failure of crops forced the Indians to slaughter their sheep for food." BIA officials became increasingly concerned about range conditions and the potentially negative effects of large herds of Navajo livestock might have on the deteriorating reservation landscape.[12]

Sheep and goats at a watering hole in front of the trading post at Kayenta, Arizona, on the Navajo reservation, 1929. Wetherill Family Collection, Arizona State Museum, University of Arizona.

In response, the BIA initiated a series of studies to assess landscape conditions and determine the precise scope of the problem. One of the most influential of these studies was produced by BIA forester William Zeh. His report provides a comprehensive assessment of the region and its conditions: water sources were scarce, and the landscape was dry and deteriorating. At the same time, Zeh is also quite critical of Navajo livestock, which he blames for the depleted conditions. "Practically all the sheep are long legged, long necked [and] inbred," he noted, adding that these animals give little wool and bear only small offspring, making them less valuable. Zeh's opinion of Navajo livestock was in line with BIA thinking, as the agency had already been working for decades to upgrade Navajo sheep from their perceived poor state. What emerges in Zeh's report, however, is an undercurrent of blame directed specifically at Navajo goat population. "An even more important factor directly responsible for the poor conditions of the range is the large number of old wethers and nondescript goats found in practically every herd." Zeh criticizes goats for depriving sheep of food necessary to survive. Ultimately, as a result of this and other studies of range conditions, the BIA began to hint at the need to reduce the total number of livestock grazing reservation lands.[13]

While much has been written about what eventually became a disastrous period in Navajo history, what I want to foreground is the increasingly negative sentiment specifically directed against goats that became part of the livestock reduction conversation of this period. While the BIA had always been critical of the Navajo's "unimproved" animals, now negative and increasingly vitriolic sentiments focused on goats as a central cause of environmental degradation. The sentiments attacked the very heart of Navajo existence.

As we've already seen, the BIA held goats in low esteem. By their own admission, the Navajo called BIA officials in the Southern Navajo Jurisdiction "goat haters" because of their negative perceptions of the animals. The former forest and range supervisor of the Navajo reservation, Virgil D. Smith, may have been one of those goat haters; in a retrospective article about his career, he remembered that the Navajo called him "Old Goat Hater" because of his attitude toward goats. Presumably Smith was not shy about expressing his opinions about the animals. Joseph Howell Jr., a forest supervisor on the Hopi reservation during the 1930s, joined in the criticism of Navajo grazing practices. According to Howell, the Navajo did not understand proper herding practices. He portrayed Navajo goats as entirely unnecessary, adding that, "the goat destroys the range that the sheep could use to better advantage."[14]

As the BIA's plans to address the deteriorating range conditions in the Southwest took shape over the course of the 1920s and early '30s, the destructive goat subplot came into increasingly sharper focus. In 1931, the US Senate Committee on Indian Affairs traveled to Southwest to conduct hearings on range conditions in the region. Over a period of several weeks, the subcommittee held meetings across New Mexico and Arizona, some within the borders of the Navajo reservation itself. While the discussions were wide ranging, goats and their perceived destructive effects on the landscape repeatedly surfaced as a pointed topic of conversation. Among the most consistently vocal sources of anti-goat rhetoric was the committee chair, Senator Burton Wheeler of Montana. Throughout the course of the hearings, Wheeler questioned Navajo shepherds about the number of goats in their care and their uses. In one representative exchange, Senator Wheeler questions a Navajo shepherd, who does not speak English, about the number of goats he keeps:

Senator Wheeler: Ask him how many goats he has.
Interpreter: One Hundred and Fifty Goats.
Senator Wheeler: Ask him what he does with so many goats.
Interpreter: He eats them, and he drinks the milk.
Senator Wheeler: Tell him he cannot drink the milk of 150 goats.
Interpreter: In order that he might save his sheep, he butchers and he eats the goats in place of killing off his sheep.
Senator Wheeler: Yes, but you tell him he should not have 150 goats. He ought to sell off a lot of the goats and get more sheep.

These are just a few among dozens of Wheeler's comments. Throughout the course of the hearings, Navajo shepherds detail the many ways they and their families keep and maintain goats while Wheeler is mostly dismissive of their statements. During one exchange, Wheeler mentions that he'd like to try some goat meat, because "I want to see how good the Indian's judgment is as to goat meat."[15]

Senator Wheeler's sentiments were not lost on the Navajo. On the day of the subcommittee hearings in Ganado, Arizona (a small town on the Navajo reservation), Native shepherds organized what might be called a goat demonstration fair which the Senate Committee members attended. A newspaper account of the occasion mentions that the senators observed a goat being milked and were invited to taste the milk. Also on display were a variety of products derived from the goat, inducing hides, leather, and rugs woven with mohair. There were no reports of the senators' reactions to the goat event, however.[16]

What's particularly interesting about the developing wave of goat panic associated with range conditions in the American Southwest during the early twentieth century is that the destructive goat narrative was not at all new. Similar rhetoric, pitting goats and goat herding against concepts of "modern" agriculture and progress, have been employed in Western culture for hundreds of years, particularly in the Mediterranean region, where sheep and goats have been herded for centuries. As early as the fourteenth century, in France, for example, officials placed restrictions on goat grazing because of their perceived destruction of landscapes, and French officials continued to wage war against peasants and their goats well into the eighteenth century, particularly after the French Revolution. In essence, anti-goat sentiment was deployed to establish a hierarchy of values that justified official control of the landscape at the expense of the poorer classes.

While the individual details of the destructive goat narrative vary according to regional context, the central components are the same: a government or other entity in a position of power seizes control of land occupied by a poorer class of shepherds whose pastoralist traditions include goats for food and sustenance. The population's use and reliance on livestock, and goats in particular, is blamed for wholesale environmental destruction. The ruling power's analysis of conditions inevitably shows that the landscape has been irretrievably harmed or destroyed due to the lack of understanding and insight of the community of shepherds, who let their goats wreak havoc.[17]

Interestingly enough, even as discussions about the deteriorating landscape within the Navajo reservation emerged in the early decades of the twentieth century, the destructive goat narrative was having a renaissance across the globe. In Morocco, a French protectorate during the early twentieth century, laws restricting pastoralist grazing contributed to the eventual dispossession of Moroccan land and livelihoods from native shepherds. In Kenya, a British protectorate during the same period, a similar displacement process was also underway. Among the key critics of Kenyan grazing practices was Sir Alfred Daniel Hall, a British agriculturalist who served as chair of the Kenya Agricultural Commission. His book, *The Improvement of Native Agriculture in Relation to Population and Public Health*, is a profoundly imperialist takedown of Kenyan culture and agriculture. "We have to save [the Kenyans] from themselves," he notes in the introduction. Hall is critical of a wide range of Kenyan livestock practices, but reserves particular ire for goats:

> Of all the livestock [kept by the native people] the goats are the worst offenders; they graze more closely on bushes as well as on grass, thereby never allowing forest growth to regenerate . . . Within historic times [goats] have been the chief agents in the deforestation of the Mediterranean, whereby hillsides have been bared down to the rock and the lower reaches of rivers choked with silt and converted into swamps.

Hall's rhetoric is eerily familiar. There is evidence that BIA officials were following events in Kenya; an internal BIA publication, *Indians at Work*, cites an article about Hall, noting "[t]his piece [about Kenya] is particularly interesting to the Indian Service as it deals with erosion as the result of overgrazing." From this perspective, BIA environmental stewardship reads like an attack on the Navajo way of life.[18]

As discussed in chapter 1, anti-goat sentiment was already a familiar story across the greater United States. During the nineteenth and early twentieth centuries in particular, Americans associated goats with the poor and immigrants, many of whom were concentrated in crowded cities. Goats and goat keeping were considered potent signifiers of backwardness, poverty, and failure to adapt to life in America. By invoking the already familiar anti-goat trope, the BIA cast its policies as natural and inevitable; anyone who was on the side of the goats, or the people keeping them, was aligning themselves against science and progress. Navajo arguments about the importance and traditional uses of goats reinforced the government's (and the national audience's) negative perspective of both the people and conditions in the region. In the hands of legislators like Senator Wheeler who had the power to create policy, the destructive goat narrative would have significant ramifications for the Navajo people.

❖

In 1933 BIA head John Collier informed the Navajo Tribal Council that the number of livestock grazing within reservation boundaries would need to be significantly reduced. Collier offered incentives for the tribe's compliance with the agency's population targets, including an expansion of reservation boundaries, along with infrastructure improvements like irrigation projects and school programs. The reduction program, eventually approved by the tribal council, was initially to be conducted on a voluntary basis, and the Navajo were to be paid for the livestock they gave up. Even so, the first round of reductions proved only marginally successful in reducing the overall livestock population: just over 86,000 sheep were removed from the reservation, many of them animals considered to be surplus or otherwise unfit. Seeing that so little progress had been made, Collier subsequently convinced the Navajo Tribal Council to sign off on a second period of reductions, with a goal of culling 150,000 goats along with an additional 50,00 sheep from the reservation. This second wave of livestock reductions became known as the "goat reduction" because of its emphasis on culling the goat population.[19]

The effects of the BIA's livestock reduction program remain seared in the minds of the Navajo people. From a logistical perspective, the reduction program proved difficult to administer, and the long distances and difficult terrain hampered officials' efforts to reach shepherds and round up livestock. From the Navajo perspective, BIA tactics were coercive and authoritarian. In some cases, Navajo tribal police appeared at people's homes demanding that shepherds give up their animals on the spot, tactics painfully reminiscent of those used to round up Navajo children and force them to attend boarding schools. Some Navajo people hid to avoid the police, while others drove their flocks deep in the hills and canyons to avoid capture. Some were so traumatized that they believed the reduction efforts were a precursor to another Long Walk, which would remove them from their homeland permanently.[20]

Accounts of the period are a testament to the devastation felt by those who lived through the era. "People kept bringing in their goats," said Navajo shepherd Ernest Nelson. "The government didn't know what to do with all those goats; so they drove them down into a Canyon near Navajo mountain and shot them in masses . . . coyotes, crows and buzzards were the only ones to have a feast." Another shepherd recalled that goats were rounded up and shot on the spot. "There was blood running everywhere in the corral as we just stood and watched." While some of the goats were sold to livestock buyers or butchered and the meat given back to the Navajo people, many goats were simply slaughtered and left to die or corralled and starved to death. Piles of animal bones were left in remote areas of the reservation, relics of the reduction era.[21]

Livestock reduction efforts continued into the mid-1940s, when the BIA's overall population targets were essentially achieved. By 1945, the total population of sheep and goats on the Navajo reservation was a combined 316,500; of that total just over 10 percent were goats. The total population was considerably less than totals of over one million throughout the 1920s and '30s. In the years following, the BIA instituted a system of livestock management centered on a permitting process. Officials determined what they termed a "carrying capacity" for any given section of land, signifying its ability to support grazing, and shepherds were issued permits for a specific number of livestock that was allowed on any given section of land based on that capacity. The grazing permit system continues as a permanent feature of range management on the Navajo reservation.

The livestock reduction period initiated an enormous economic shift on the Navajo people. Navajo shepherd Jim Counselor described the state of the reservation in 1936: "I can show . . . that 70 or more Navajo families out here who have an average of six sheep and goats left. I can show [you] other families who haven't any stock left, not even horses." New Mexico Representative Antonio Fernández testified before the US House of Representatives in 1949 about the dire conditions in the years following the reductions, noting that some on the reservation had been forced to turn to hunting wild jackrabbits for food, derisively calling the rabbits "Collier goats." The name "John Collier," used in any context, is still likely to raise the ire of contemporary Navajo.[22]

Experts past and present generally agree that the landscape of the Southwest was severely depleted by the early twentieth century. Some form of livestock reduction was undoubtedly necessary on the Navajo reservation (and in other parts of the Southwest) to arrest the deterioration of the severely taxed landscape. And it's also true that many Navajo herders were well aware of the deterioration of the landscape and the need for reducing livestock numbers. Nevertheless, what was missing from the livestock reduction conversation of the 1930s was a comprehensive analysis of reservation conditions that considered every factor that contributed to the deteriorating landscape. Among them was a more holistic analysis of the contributions of all breeds of livestock toward landscape conditions in the region. BIA-imposed improved livestock taxed the landscape more than Navajo breeds, which were well adapted to range conditions; not to mention the parallel effects of extensive overgrazing by Anglo sheep and cattle ranchers in the region during the 1880s and '90s. The BIA's characterization of Navajo goats as a central culprit of widespread landscape destruction and erosion was one key means by which the agency evaded the nuances of an exceedingly complex environmental and social problem that the bureau was ultimately ill-equipped to handle.[23]

Navajo Goats in the Twenty-First Century

In 1947, Navajo Tribal Council member David Clah voiced a startling sentiment: "We all know that in the past we lived from sheep, but the sheep is a thing of the past now and our future generations are growing up without any sheep." His words would prove prescient. In 1978, Navajo leaders declared that, for the first time, the number of Navajo who earned their living wages off the reservation surpassed the number of those making a living raising livestock. The *Navajo Times* newspaper lamented, "[p]eople say that the shepherding profession on the Navajo Reservation is slowly dying."[24]

From a present-day perspective, it's clear that Navajo shepherding has not disappeared, though the practice has changed. Most significantly, sheep and goats no longer represent a primary source of livelihood and subsistence. The BIA-administered grazing permit system, first implemented in the 1930s, continues to regulate reservation grazing practices. The number of permits is limited, restricting access to traditional grazing grounds. Those without permits can still keep sheep and goats but must keep them within a pen or other confined area, and owners must rely on hay and other feed for the animals, which can be expensive. Even so, some Navajo graze sheep and goats across the landscape in direct defiance of the permit system, leading to conflicts over the land and its uses. Over the years since the system was implemented, access to grazing permits has also been hampered by the process of transferring permits from generation to generation, particularly when permit ownership becomes tied up in probate after the death of a holder. More recently, new permit holders are required to complete a conservation plan in cooperation with the BIA, adding another layer of bureaucracy to contemporary shepherding.[25]

Some see the Navajo National Trust Land Leasing Act of 2000 as a potential avenue toward restructuring and reviving Navajo shepherding practices. Among other things, the act gives the Navajo Nation authority to lease tribal lands for agricultural purposes, though the lands are still held in trust by the BIA. In this way the Navajo Nation might, at minimum, gain more control over the land within its borders, enabling it to ease restrictions that have hampered shepherds and prospective shepherds for decades. Speaking about the act at the Sheep is Life conference in 2023, W. Mike Halona, director of the Navajo Department of Natural Resources, said, "we need to take a look at our own land management to benefit producers and preserve our way of life. . . . the younger generation coming up want to raise livestock and aren't able to get a grazing permit. Our hope is that in fifty years, we will still have sheep raisers [among us]." Increased access and control will enable the Navajo to restore the balance between people, land, and livestock disrupted during the livestock reduction period.

❖

In 1981, Navajo Tribal Chairman Peter McDonald discussed the changes in the Navajo diet over the course of the tribe's history. "Traditionally our people ate well," he wrote in the *Navajo Times*. "We hunted or grew what we ate . . . we did not worry about the effects of preservatives or cancer causing hormone injected meat or chemically derived food additives or coloring." McDonald noted the place goat's milk once occupied within Navajo food traditions. "Prior to the 1930s goat's milk and cheese were an essential ingredient of our diet, but the goats were the first to go [during the livestock reduction period] . . . and calcium rich foods ceased to be a part of Navajo life." McDonald called on the Navajo to look to their history for solutions. "We used to grow our own corn," he wrote, "we can do it again . . . we can keep a flock of chickens, raise a few animals for meat, and work toward self-sufficiency."[26]

Among its many effects, the decline of shepherding has had a significant impact on Navajo nutrition. Modern Navajo diets are mostly devoid of the foods of their ancestors: yucca root, prickly pear, or cornmeal with warm goat's milk. It's no secret that rates of obesity, diabetes, and other diseases are far higher among Indigenous populations, including the Navajo, than in the general United States population. Additionally, a 2014 study found that 76.4 percent of individuals on the Navajo reservation experienced some form of food insecurity. What Winona LaDuke has called the "ravages of colonial food" have had profound effects on the health of the population.[27]

The Navajo are working hard to reclaim control over their traditional food sources. Through the efforts of community health advocates like Denisa Livingston of the Diné Community Advocacy Alliance, the Navajo Nation passed the Healthy Diné Nation Act in 2014. The law instituted a 2 percent tax on unhealthy foods like sodas and junk food with low nutritional value sold on the reservation. Despite some resistance, as of 2020, the tax had raised over nine million dollars of revenue, which have been redistributed to chapters across the reservation. Livingston also focuses on a number of other food initiatives, including taste education, introducing young Navajo to traditional foods like meat from Churro sheep, prickly pear cactus, and piñon nuts. A study of the Navajo Nation food system conducted by the Diné Policy Institute maintained that food sovereignty is part education and part spiritual journey: "Diné food sovereignty is fundamentally achieved by upholding our sacred responsibility to nurture healthy interdependent relationships with the land, plants and animals that provide us with food."[28]

Efforts to rebuild Navajo food systems face a familiar opponent. The returning nightmare of drought continues to challenge Diné agriculture and livestock. During the 1960s, conditions were bad enough that the Navajo Nation instituted its own programs to reduce the numbers of livestock within reservation boundaries; another severe period of drought during the 1990s prompted a similar initiative. Unfortunately, drought conditions are ongoing in the region: the period from 2000–21 was the driest period in over 1,200 years. The creeping effects of worldwide climate change are visible within the boundaries of the Navajo Nation. One study analyzing the progression of dune formation in the western part of the reservation determined that during the period between 2009–10, the dune fields grew by over 229 feet. The continued deterioration of arable landscape represents an ongoing challenge to the Navajo way of life.[29]

Navajo shepherding traditions continue to evolve in the face of change. Despite myriad rules and regulations and grazing permit requirements, the livestock population numbers within the boundaries of the Navajo Nation remain substantial; according to US Department of Agriculture census figures, there were 194,000 sheep and 68,000 goats on the Navajo reservation as of 2017—there are more goats residing on the Navajo Nation than in most states. The contemporary goat population of Navajo Nation consists of a wide cross section of breeds, including Angora goats used for both fiber and meat, and Boer goats kept primarily for meat purposes. In addition, some Navajo also keep dairy stock like Alpine and Nubian goats. Other keep goats alongside traditional Navajo Churro sheep and herd them as their ancestors might have. That being said, according to several regional livestock experts I spoke with, the more recent trend within the Navajo Nation is toward cattle raising, which brings a higher financial return.

During the early twentieth century, wool and mohair sales were one significant source of income for the Navajo. While the market was strong then, wool and mohair bring only pennies per pound today. Despite this, Navajo weavers continue to use wool and mohair from their livestock in creating rugs, saddle blankets, clothing, and other traditional textiles. Weavers have individual preferences when it comes to the material they prefer for weaving purposes; some prefer Churro sheep wool, others Merino or Rambouillet wool, while others prefer mohair. "The wool of Angora goat takes the native dyes very well," one Diné weaver told me, "A small amount of sheep wool and Angora goat wool carded together can make a good combination." Some weavers use mohair specifically for the warp, the vertical strands through which strands of wool yarn are drawn to form a rug; the mohair serves as the support structure as the weaver constructs the textile. The strength of mohair fiber makes it more suitable for this purpose than wool. Weaving remains a means through which Navajo traditions are passed into the future.[30]

Educators like Diné weaver and fiber artist Roy Kady are actively working to teach a new generation of Navajo about sheep and goats. Kady is a board member of the Sheep is Life organization, a nonprofit that works to restore Navajo culture, life, and land. He's also a shepherd who keeps a small herd that numbers around thirty-six sheep and goats. Kady has a particular affection for goats: "My people were goat people," he says, who preferred goat meat to sheep meat. Kady says his mother and grandmother passed their shepherding skills to him. For Kady, goats are integral to shepherding practices. "Goats are the ones that have a sense of place and time," he said. "They are a lot more intellectual than sheep. When you are out grazing, the goats are the ones that lead the flock back home. Goats look around. The sheep just move forward, munching, heads to the ground." During the summer months, Kady is likely to be found at sheep camp with his flock, grazing in remote areas where the range is lush. Navajo students and artisans apprentice with Kady, who teaches them the ins and outs of shepherding and livestock care. Contemporary Navajo shepherds, weavers, and activists are all actively working to ensure that future generations will sustain their ancestral traditions.[31]

❖

For centuries the Navajo have contended with shifting narratives of assimilation and progress dictated by outsiders including Spanish colonists and the United States government. While sheep and goats once sustained the Navajo people, the United States government perceived Navajo livestock as deficient. After US authorities established the Navajo reservation in 1868, it worked to remake Navajo livestock in its own image, introducing outside breeds with the goal of improvement, or conformity, with Western standards of economic productivity.

The livestock reduction period of the 1930s marked a low point in Navajo history. But the story of Navajo sheep and goats does not end there. The Navajo have always remained resilient in the face of change; the deep spiritual link between the Navajo, their land, and their sheep and goats continues. Many Navajo shepherds and families continue to keep small numbers of sheep and goats for personal consumption or for ceremonial purposes, practices which keep traditions alive. Navajo leaders and advocates remain resolute as they work toward a future of self-determination, and a restoration of *hózhó*, the fundamental balance of people, land, and livestock. In an oral history discussing the evolution of Navajo history during his lifetime, Diné elder Hoke Denetsosie remarked: "Our adaptation to transition from the old life to the new one should help our people in the future. In the final analysis it is our Navajo heritage and the significance of our Navajo beliefs which bind our people—the Diné—together[.]"[32]

OPPOSITE: *Ode to Goats* by Cormac B. Fawsitt (2021) at Ayers Brook Goat Dairy in Randolph, Vermont. Used with permission.

II

Commercial Success With Dairy Goats

CHAPTER THREE

How Goat's Milk Became Healthy

Most good things are ridiculed before they are appreciated. [The goat] is no longer a joke, and now deserves to be placed in a class with other four footed aristocrats of the animal kingdom.

Chicago Tribune, 1919

Americans typically assume that the fluid substance called "milk" is cow's milk. Although other mammals produce milk that is regularly consumed by humans, including goats and sheep, cow's milk has become so commonplace that it has assumed a role as the generic concept of milk itself. But cow's milk's present-day dominance in the consumer marketplace is a relatively new historical development. The rise of cow's milk represents over a century of evolution in American consumption habits alongside advances in scientific knowledge about microorganisms and their role in food safety. There was actually a time when goat's milk was a significant contender as the American public's milk of choice.

In order to trace the winding path that led to the widespread popularity of goat's milk in the United States, we have to start in an unlikely place—a distillery. For centuries distilled spirits were produced in small-scale operations on or near farms that grew the grain used in the distilling process. During the nineteenth century, the Industrial Revolution transformed the craft with the invention of a new type of manufacturing apparatus called the column still, a revolutionary device that enabled distilleries to produce alcohol continuously instead of by the older and more labor-intensive single-batch method. The innovative production method paved the way to the era of the mass production of liquor, and large-scale distilleries moved into fast-growing cities across the United States.

More importantly for our purposes, this new generation of distilleries and breweries generated a secondary phenomenon that became known as the distillery dairy. This odd-sounding institution came about when urban distillery owners built barns near or adjacent to their facilities where they housed a variety of livestock, including dairy cows, which consumed the spent grains from the distilling or brewing process. In other cases, cattle feedlots paid distilleries for their waste, which was transported to urban dairy operations. The end result was that milk from cows fed on distillery waste was sold to city residents. The arrangement allowed owners of distilleries and breweries to generate a profit from what was otherwise a waste product.

The development of this urban dairying system was symptomatic of the nation's fast-changing food landscape. Rapidly growing nineteenth-century cities significantly changed the relationship between people and their food. Farms, which required land to grow crops and raise livestock, became increasingly distant from population centers. The distillery dairy solved the problem of increasingly remote sources of milk, which was hard to transport and keep fresh, by providing a more conveniently located supply. Across the country, from New York to Philadelphia to Baltimore on the East Coast and in other fast-growing cities such as Cleveland, Milwaukee, and Chicago, the distillery dairying system became the primary source of milk for urban residents.

Meanwhile, conditions inside distillery dairies were awful. While spent grains left over from the brewing and distilling process have some value as animal feed, this was typically the only food the animals received, dumped hot from the still. Stable conditions were horrific as cows were typically packed tightly together in stalls with no room to move. Many developed ulcerated sores, and the accumulating filth often damaged cows' hooves so badly that many were unable to stand. Month after month of a ration of what was termed "still slops" rotted cow's teeth. It was not unusual for cows to become so weak and sick that workers hoisted the animals up on mechanical lifts in order to milk them.

Not surprisingly, the milk produced at these urban dairies was revolting in both appearance and taste. Often described as bluish and watery in appearance, one particularly vivid account described swill milk vividly as "chalky, cheesy, dirty slop," which one drank "with averted eyes, talking zealously [all] the while of other things." In order to divert consumers' attention from such poor-quality milk, dealers typically employed a variety of means to doctor the end product. Many added water to increase the volume of milk and thus the resulting profit; the practice was so common that metal water pumps became known as "iron cows." Other nefarious milk dealers added substances such eggs, starch, or plaster of paris to make the milk seem creamier; molasses might also be added to improve the appearance. To make matters worse, dishonest dealers often labeled such concoctions as "pure country milk."[1]

Robert Hartley, a New York City businessman turned social reformer, became one of the most prominent voices in an emerging campaign against distillery dairy practices. In a series of articles published in the *New York Observer* and later in a book called *An Historical, Scientific and Practical Essay on Milk*, published in 1842, Hartley systematically laid out the case against poor-quality milk. "The manner of producing milk to supply the inhabitants of cities," said Hartley, "is so contrary to our knowledge of the laws which govern the animal economy that from a bare statement of facts, any intelligent mind might anticipate the evils which actually result from it." Most disturbing were the effects of such milk on human health, especially the very young, as infant-mortality rates in New York and other cities had skyrocketed. Rooted in the emerging temperance movement's critique of the immoral nature and corrupting influence of alcohol, Hartley maintained that alcohol interests had significantly corrupted milk from its once pure, pastoral state, and the effects were dire. "[B]y basely counterfeiting an article of food and imposing in on the unsuspecting ... health is deranged, lingering and distressing diseases are introduced, and life itself is destroyed with impunity."[2]

Hartley's efforts provoked a swell of outrage against distillery dairies in New York City. In 1858, *Frank Leslie's Illustrated Newspaper*, a popular tabloid of the period, took up the cause of milk reform and in doing so exposed the practices to a wider audience. "For the midnight assassin we have the rope and the gallows; for the robber the penitentiary; but for those who murder our children by the thousands [the distillery dairies and swill milk dealers] we have neither reprobation nor punishment," wrote the publication in a grim tone. Even more persuasive were the etchings of distillery dairies published by the paper, the paparazzi photos of their day, which provided disturbing visual evidence of distillery dairy practices. A swill milk war began in New York City, and spread to cities across the country as concerned citizens began to take action.[3]

From Bad to Worse: Tuberculosis Milk

Meanwhile, everyday life during the nineteenth century was clouded by the lurking presence of an epidemic disease known as consumption. If you didn't have consumption yourself, you probably knew someone who did—or you just didn't have symptoms yet. The disease was relentless, attacking anyone regardless of cultural background, religion, or economic status. Even as recently as 1900, consumption, which later became known as tuberculosis, was the leading cause of death in the United States.

Although consumption was widespread in Europe and the United States during the nineteenth century, the disease has actually existed for thousands of years. Scientists have identified the tuberculosis bacterium in Egyptian mummies through modern DNA analysis; the *Huangdi Neijing*, an ancient Chinese medical text from 400 BCE describes a wasting sickness similar to tuberculosis. Ancient Greek physician Hippocrates, writing in the fifth century BCE, described the disease and its characteristic fever and coughs; he called the illness phthisis. Over the course of millennia epidemics waxed and waned, only to reappear with a vengeance in the industrializing regions of the globe during the eighteenth and nineteenth centuries, killing millions.[4]

Despite the enduring presence of the disease, human understanding of consumption had essentially remained static for centuries. Among the earliest treatments included bloodletting, a therapeutic practice that originated in ancient Egypt but continued well into the nineteenth century. The practice of bloodletting was grounded in the belief that the human body was controlled by four humors (blood, phlegm, black bile, and yellow bile); removing blood from a patient's body was considered a means of rebalancing the humoral system and thus restoring health.

An 1869 article in *The Atlantic Monthly* detailed another contemporary theory of disease: "Wetness of the soil is a cause of [consumption] to the population living on it." The article describes two brothers, one who lived in a house "bathed in sunlight" and on well-drained soil. The other brother lived nearby, but his house did not receive as much sun and was susceptible to "damp and chilling emanations [that] arose from the meadow." The family living in the sunny house remained unscathed by consumption, while the latter did not. The idea that clouds of disease vapors emanating from soil or decomposing matter could cause disease, known as the miasma theory, was, like the theory of the four humors, still widely believed well into the nineteenth century. Unhealthy miasmas were thought to be the source of a wide variety of diseases, including cholera and the bubonic plague, as well as consumption. During the nineteenth century, consumption even gained a reputation as a vaguely romantic affliction. Writers and artists such as poet John Keats, the Brontë sisters, and Henry David Thoreau, suffered from the disease, fueling the notion that consumption affected primarily the young and creative.[5]

These and many other entrenched beliefs about the nature and causes of consumption were finally put to rest when German physician Robert

Koch identified the cause of the disease, a bacterium that he named *mycobacterium tuberculosis* (*M. tuberculosis*) after the nodules, or "tubercules," commonly found in the lungs of those with the disease. Koch's 1882 discovery marked the transformation of the illness called consumption into the more ominous sounding tuberculosis, a modern disease caused by a dangerous lurking microorganism invisible to the naked eye. Not only was Koch's discovery a significant milestone in the scientific understanding of consumption, but it marked a key moment in the advancement of the modern germ theory of disease.

Although Koch received a Nobel Prize for his work in 1905, the identification of the specific *cause* of tuberculosis did not translate into an immediate *cure*. It was not until the 1940s that scientists found the first effective treatment for tuberculosis, streptomycin. But even that victory proved relatively short-lived; in recent years antibiotic resistant strains of *M. tuberculosis* have emerged. Ultimately, an absolute cure for tuberculosis has proven to be an infinitely receding target.

Once the cause of tuberculosis was identified, physicians and scientists rushed to find a way to stop the disease. In 1890, Robert Koch announced the discovery of what he called tuberculin, a substance created from tuberculosis-infected cells. While tuberculin was quickly discredited as a cure, scientists noticed that tuberculin was useful as reliable *test* for tuberculosis, since those administered with the substance produced a consistent and identifiable skin reaction. The emergence of a test for the disease was a significant advance, as it enabled doctors to identify those infected with tuberculosis and to initiate a variety of measures to limit the spread of known infections. The tuberculin test would later prove useful when public health officials were forced to address the presence of *M. tuberculosis* in the nation's milk supply.

❖

While the precise cause of consumption remained mysterious for millennia, humans had long observed that animals suffered from a similar disease. In the first century CE, Roman agricultural writer Columella noted symptoms in cattle, including weight loss, coughing, and ulcerated lungs, which culminated in phthisis (Hippocrates' term for tuberculosis). During the nineteenth century, many physicians and scientists believed that the same mechanism was responsible for tuberculosis in both humans and cattle, and that the cattle disease presented no threat to the human population. Soon, however, scientists identified an entirely separate strain of the tuberculosis bacterium, *Mycobacterium bovis* (*M. bovis*).

Over the ensuing decades, evidence that diseased cattle might be capable of infecting humans began to emerge. In 1883, the American Veterinary Congress recommended against consuming milk from cows that could be shown to have tuberculosis. In 1899, Harvard University issued a study that proved that cattle could transmit tuberculosis through milk. By the turn of the twentieth century, it was clear that the *M. bovis* strain of tuberculosis was present in the milk of infected cows and could be passed to, and infect, humans. The implications were dire: the nation's milk supply was potentially contaminated, even deadly. Some experts estimate that as many as 20–30 percent of human tuberculosis infections during the nineteenth and early twentieth century were probably from *M. bovis*.[6]

The already widespread conversation about poor-quality, contaminated milk from distillery dairies was reenergized by this newly defined public health threat. Revelations about the existence of *M. bovis*, and the potential for its transmission through cow's milk and meat, provided a hard focus for the general public. Dramatic newspaper headlines grimly proclaimed "Death Lurks in Milk," a sentiment that was repeated across the country for decades. A Nashville physician expressed the prevailing mood of the populace: "one of the worst evils to be put out of Nashville is diseased milk—milk with death in it. Every day . . . milk is farmed out to customers which comes from Tuberculosis cows." Efforts directed toward cleaning up or eradicating distillery dairies were not going to be enough to ensure milk safety; this newly identified microscopic contaminant required an entirely different approach.[7]

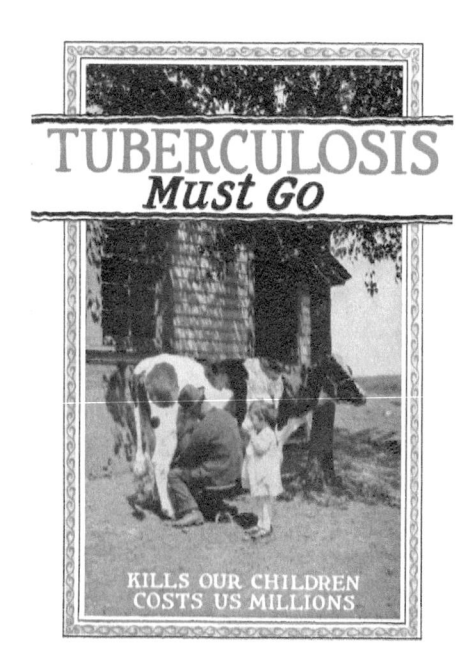

Meanwhile, milk safety advocates pushed forward with strategies for cleaning up the nation's milk supply. Pasteurization, a heat-treatment process named after its developer,

WHI-33959, Collection of Wisconsin Historical Society.

French chemist Louis Pasteur, was fast becoming a growing part of the contemporary milk safety conversation. While the process of heating milk to kill harmful microorganisms appeared promising, some considered the process a new but unproven technology. Many scientists, physicians, and pediatricians of the period maintained that the application of heat was potentially counterproductive and could destroy milk's nutritional properties altogether. Others felt that pasteurization could have the effect of masking dirty milk and poor handling practices, arguing that its use could lead to even more abuses. In an era of deep suspicion about milk adulteration, pasteurization appeared to some as another potentially negative intervention.

New Jersey physician Henry Coit envisioned a different means of cleaning up the milk supply: a system of inspecting and certifying milk for human consumption. Coit's son had died from drinking contaminated cow's milk, and so the physician had considerable interest in ensuring the safety of the milk supply. Coit advocated for the formation of private milk commissions across the country charged with administering a milk certification process. Under the system, dairies would be subject to a variety of requirements, including inspection of their facilities and the monitoring of animal care and milk-handling practices, and their milk would also be tested regularly for contamination. Assuming a dairy met and passed all of the requirements, its milk would be allowed to carry a "certified" seal, assuring consumers of its quality. While the certified-milk system was adopted in over forty cities across the country, there were a number of barriers to its full acceptance. The costs of administration were high, and as a result the price of certified milk was often at least double the price of regular milk, making it unaffordable for many families.

Others developed a global initiative to attack the tuberculosis milk problem at its source: cattle infected with *M. bovis*. The United States Department of Agriculture (USDA) issued stern directives to farmers regarding the potentially deadly effects of poorly maintained herds and unsanitary milk-handling practices. In a 1911 farmer's bulletin, the agency warned, "Milk is the staple food of infants and young children . . . if the milk is from a tuberculous cow, it may contain millions of living tubercle germs. Young children fed on such milk often contract the disease, and it is a frequent cause of death among them." Using the tuberculin test developed by Robert Koch, public health officials from all over the world began a campaign to identify cows with tuberculosis and eliminate them as a means of limiting the spread of additional tuberculosis infections. Finland became the first country in the world to undertake a bovine eradication program with the express purpose of culling animals that tested positive for tuberculosis in the 1890s, and other countries soon followed. In 1917, the United States Bureau of Animal Industry (BAI) started its own campaign known as the National Tuberculosis Eradication Program. Over the next several decades some 3.8 million out of a total population of over 66 million cattle in the United States tested positive for tuberculosis and were killed. While many dairy farmers across the country found the program's efforts draconian and resisted, by 1940 tuberculosis had been all but eliminated from the nation's cattle population. To this day, the National Tuberculosis Eradication Program continues to monitor the presence of tuberculosis in the nation's cattle population.[8]

First there was swill milk, then tainted and adulterated milk. By the turn of the twentieth century, yet another type of contaminated milk had emerged: tuberculosis milk. What was a concerned citizen to do? Amidst the growing furor over diseased cattle and the widespread implications for public health, the humble domestic goat began to e\merge as a potential solution.

Goats as Milk Saviors

The discovery of microorganisms capable of causing human illness and disease revolutionized the medical profession. The germ theory of disease enabled human knowledge to finally advance beyond ancient theories of illness such as the presence of harmful vapors or internal humoral imbalances. The era's scientists and physicians rushed to understand these newly discovered microscopic invaders that were powerful enough to attack and even kill living beings. Even so, it would take decades to sort out the precise mechanisms by which microorganisms such as *M. tuberculosis* infected and sickened humans and to develop effective and reliable ways to stop them.

Even before Robert Koch made headlines with the discovery of the *M. tuberculosis* bacterium in 1882, European scientists had been studying the mechanisms by which diseases of all kinds were passed between humans. In one of the more significant of those early experiments, French scientist Jean-Antoine Villemin conducted an experiment in which he injected a variety of animals, including rabbits and goats, with tuberculosis-infected tissue in 1865. Though rabbits and other animals subsequently came down with the disease, others—including goats—did not. While Villemin's experiments established the infectious nature of tuberculosis years before Koch identified the specific mechanism of transmission, the fact that some animal species came out unscathed in Villemin's experiments did not go unnoticed, and the question of the humble goat's immunity to tuberculosis developed into a serious public health issue. The subsequent push to determine the potential immune properties of goats was one of many ways physicians and scientists attempted to bridge the nagging void between the cause of consumption and its cure.[9]

As scientists investigated the physiology of goats, some found what they considered to be proof of the animals' immunity. One of the most often cited pieces of empirical evidence in this regard was related by French scientist Edmond Nocard, who observed that among 130,000 goat carcasses examined in Paris slaughterhouses, he and his colleagues observed no cases of tuberculosis among them. Similar observations at slaughterhouses in Belgium and other European countries corroborated the French conclusions. Carl G. Wilson, a physician and prominent goat's milk advocate of this period, wrote in *The Goat World* magazine that for the period between 1904 and 1914 the BAI reported that "not a single goat was [found to carry] tuberculosis."[10]

Researchers became determined to understand the internal mechanisms of goats' apparent immunity. One of the key areas of research was the quality and nature of goat's blood. For millennia, bloodletting, or the process of taking blood from the body, had been used as a means of restoring human health. Starting in the seventeenth century, European physicians had started exploring the opposite concept: transfusion, or the process of *adding* blood to the body. Transfusion experiments using both human and animal blood had become increasingly popular throughout the course of the nineteenth century. Now, the prospect of transfusing blood that carried potential curative or immune properties enticed those looking to find a cure for an entrenched epidemic disease like tuberculosis.

One of the most famous of the era's goat's blood-transfusion experiments occurred in France in 1882, when French physician Samuel Bernheim attempted to prove a goat's immune properties by transfusing goat's blood directly into a woman suffering from tuberculosis. While the experiment was remarkable in and of itself, Bernheim had both personal and public health motives for his experiments. Bernheim commissioned a painting of himself performing the transfusion by prominent Paris artist Julian Adler, presumably anticipating the widespread acclaim that would result from his anticipated medical breakthrough. Bernheim promised Adler a 1,200-franc commission for creating the work and a 200-franc bonus if Adler could get the painting accepted into the prestigious Paris Salon. Outside of his self-promotion campaign, Bernheim eventually transfused a total of eleven patients with goat's blood, seven of whom he claimed to have cured completely.

Likewise, French scientists Georges Bertin and Jules Picq conducted a series of experiments that showed that goat's blood had a bactericidal effect when injected into tubercular rabbits, as many of the injected rabbits either improved or were cured. Their work was significant enough to warrant a mention in *The New York Times*, which (prematurely) declared a "New Cure for Consumption." Another French physician, Dr. Baradat of Cannes, reported similar success when treating tuberculosis patients with serum (plasma) derived from goat's blood. Baradat believed that goat serum was not only "antitoxic and bactericidal, but also that it is a 'tonic' that is to say, stimulating and regenerative."[11]

Some researchers even pursued transfusion studies using goat's milk; milk transfusion research was a less common subset of the era's blood-transfusion research. While blood was the primary focus of study, some researchers turned to milk because of the lack of consistent availability of human or animal blood, and because fresh blood quickly coagulated, rendering it useless for transfusion if not found and used immediately. The practice of milk transfusion was not widespread; the procedures were mostly unsuccessful and often killed patients. Nevertheless, as the theory of the special immune properties of goats took hold, some investigated the effect of goat's milk transfusions as well. One Irish physician conducted an experiment in which he injected goat's milk directly into tuberculosis patients. He said that although the milk transfusions did not cure the disease, he was able to prolong the lives of patients.[12]

Eventually theories regarding goats' potential immune properties gained enough scientific traction that medical authorities began to jump on the goat bandwagon. In 1898, prominent British physician William Broadbent, physician to Queen Victoria, declared outright that goats do not suffer from tuberculosis. Many physicians followed the trend, among them Finley Bell, a pediatrician affiliated with the New York Academy of Medicine, who declared in 1906 that the goat "is not subject to tuberculosis or other diseases in this climate." Editors of the *British Medical Journal* hedged their bets by playing both sides: "[G]oats, if not actually immune, are very refractory [that is, resistant] to [tuberculosis]." Even longtime milk safety advocate Milton Rosenau, a professor at Harvard Medical School, declared in his influential book *The Milk Question* that "one of the great advantages to goat's milk is that goats are practically immune to tuberculosis, and thus this danger is at once eliminated."[13]

Overall there were relatively few negative scientific pronouncements concerning goats and their immune properties; among the doubters were researchers from the relatively new field of veterinary medicine. Citing examples in which they said goats were believed to have contracted tuberculosis, veterinary researchers from Iowa State College spoke for many in their profession when they warned, "The fact is that Tuberculosis in goats is not nearly so rare as has been thought, and the false security felt by those using goat's milk because of the belief that there is no possibility of the transmission of Tuberculosis is a dangerous thing." Perhaps not surprisingly, the general public was not interested in taking health advice from veterinarians.

Although many physicians and scientists of this era came to believe in the immune properties of goats, in the end there was never any consensus on the issue. In part, this was likely because immunity itself was a new concept, the contours of which were only beginning to be studied (and even now are still not fully understood). In an era just a few years removed from centuries-old beliefs about bodily humors and the infectious properties of vapor clouds emanating from the soil, scientific certainty was really anyone's to claim.

Despite the fact that a variety of goat-related experiments conducted during this era showed somewhat promising results, momentum for the study of goats and their blood in treating tuberculosis waned by the 1920s. Contemporary medical historians cite the lack of sufficient clarity around the efficacy of the therapies, rooted in inconsistent formulations as well as the absence of controls in many of the studies. The promised "cures" were difficult to replicate and there were no clear and astounding therapeutic results. Even so, more than a century later, some are suggesting that these early experiments be revisited to further assess their potential efficacy.[14]

Medical science eventually moved on entirely from the theory that goats could save humanity from tuberculosis through their purportedly innate immune properties. But if blood wouldn't solve the problem, there was another potential health tonic generated by goats—their milk. The idea of consuming what was thought to be tuberculosis-free milk from goats emerged as another potential solution to the problem of contaminated cow's milk. Goat's milk offered an appealing and potentially safe option for those desperate to protect themselves from the dreaded disease of tuberculosis, which had killed millions of people across the world. The developing nationwide interest in goats and their milk would eventually drive a thriving commercial market for goat's milk in the United States.

There were a number of reasons society began to look toward goat's milk. Goats were not associated with distillery dairies, which produced the widely publicized filthy and unhealthy swill milk from diseased cows. Goat's milk also subverted the pasteurization debate; as the thinking went, one need not worry about whether goat's milk should be pasteurized because it did not, and could not, carry tuberculosis to begin with. Even more persuasive were the numerous positive, or at least positive-leaning, pronouncements from scientists and physicians about goat immunity that accumulated over the course of the late nineteenth and early twentieth century. As a result, goats and their milk began to acquire the pastoral, healthy glow that, as Robert Hartley had pointed out decades earlier, was long lost for cows and their milk. In an era of dozens of new and competing scientific theories about human health, the solution was simple and appealing: cow's milk was bad, goat's milk was good.[15]

And so, over time the gradual accumulation of evidence pointing toward goats and their immunity to tuberculosis fueled an improbable but increasingly positive public image of the animals. The popular press began to extol the virtues of the once maligned goat. Headlines that once raged over nuisance goats now actually encouraged the public to reconsider their previous impressions of the animals. As early as 1888 the *Brooklyn Daily Eagle* announced, "Goats Misunderstood," noting that an entrepreneur was hoping to establish a goat dairy there. "For invalids and consumptives, those who are greatly run down by disease, [goat's milk] is an excellent tonic." A 1911 article reprinted in dozens of newspapers across the country trumpeted "American Goat: A Ridiculed Hero," heralding the "elevation of the Goat to its Rightful Place as a Monarch of the Dairy." The long-entrenched negative reputation of goats began a remarkable turnaround. Once cast as nuisance animals associated with all manner of poverty and urban decay, goats reemerged in the cultural zeitgeist as extraordinary animals equipped with an innate power to save humanity.[16]

Goat's Milk as Medicine

As medical professionals began to embrace the health properties of goat's milk, it was increasingly employed as a therapy in hospitals and sanitariums. Sanitariums were specialty facilities designed to house and care for the sick, and during the nineteenth and twentieth century became synonymous with tuberculosis care. The tuberculosis sanitarium movement began in Europe and soon spread to the United States, where the first formal sanitarium developed specifically for tuberculosis patients was established at Saranac Lake, New York, in 1873. Sometimes medical authorities sent those suffering from tuberculosis to such institutions to isolate them, and sometimes sufferers, typically those with means, chose to go. Hospitals also set up tuberculosis wards resembling sanitariums where tuberculosis sufferers were housed separately from the general hospital population. The Indian Health Service established separate sanitarium facilities for Native populations across the West, while the Piedmont Sanitorium in Virginia provided one refuge for Black tuberculosis sufferers barred from most other institutions. While *M. tuberculosis* did not discriminate by class or race, humans did.[17]

Notably, and perhaps ironically, physicians have prescribed milk as a tuberculosis treatment since the days of the Roman Empire. Modern therapies varied by doctor and facility but typically advised tuberculosis patients to drink multiple quarts of milk a day. The threat of swill milk, adulterated milk, and especially tuberculosis milk complicated the administration of cow's milk in a health-care setting, especially one devoted to tuberculosis care. Institutions of the day generally took one of two approaches to the problem: some used cow's milk that was either certified or pasteurized and therefore presumed to be healthy and uncontaminated, and others turned to goat's milk. The emerging positive reputation of goats fit naturally into the sanitarium regimen, creating an aura of healthfulness and safety to the facility providing it. Goat's milk bestowed the promise of health, and even potential to cure, to what was otherwise essentially palliative or hospice care performed within the sanitarium setting.[18]

Institutional demand for goats and their milk grew quickly. One particular center of interest was in Chicago, where, in 1920, twenty-one purebred Toggenburg dairy goats arrived at the Oak Forest Tubercular Hospital. The moment was the result of several years of planning on the part of Cook County commissioners, who administered the county owned facility. While other commissioners were initially skeptical of the idea of allocating county funds for goats, Albert Novak was a particular advocate. He related in a 1919 meeting that his uncle had tuberculosis and was near death at one point. "He was put on a diet of goat's milk. He drank it four times a day warm from the goat. And he got well." At the same meeting, commissioners were invited to sample the goat's milk from a local dairy. A physician from local Michael Reese Hospital testified that a supply of goat's milk for infants in the hospital's care was an "urgent need." The superintendent of Oak Forest Hospital said that goat's milk would be "of inestimable value in the children's wards and for tubercular patients" but he could not find a supply. Eventually the idea of funding a goat farm was put up for a vote. Funds were eventually approved, and the Cook County Commission created a subcommittee called the Goat Commission, which for several years administered the goats and facilities for the county.[19]

One of the most vocal of Chicago area goat advocates was Amanda Louise Buchanan Patten, commonly referred to at the time as Mrs. James Patten. James Patten was well-known for having built a substantial fortune

The Oak Forest Tubercular Hospital in Cook County, Illinois, kept a herd of Toggenburg dairy goats as a milk source for its patients. DN-0080092, Chicago Sun-Times/Chicago Daily News Collection, Chicago History Museum.

in the commodities trade, and Mrs. Patten, like many wealthy society women of the period, was involved in a variety of charitable causes in the Chicago area. Mrs. Patten's involvement with goat dairying became a central feature of her charity efforts, and she was among the area goat advocates that served on the Cook County Goat Commission.

After her commission service, Mrs. Patten turned her attention to nearby Evanston, north of Chicago, where the Pattens had constructed a large mansion. She presented the Evanston Hospital a goat as a starter animal for the facility to start its own herd. According to reports, the goat "went on strike" and refused to give milk. Mrs. Patten responded to the story by saying, "I believe that there is nothing the matter with it except that it's lonesome," and promised to deliver another goat to the hospital. By 1922 the Evanston Hospital herd had grown to twenty-five goats; their caretaker remarked that they were also "good lawn mowers." Mrs. Patten eventually established her own commercial goat dairy, the Evanston Goat Milk Company, in 1922. According to reports Patten intended to sell the milk for fifty cents per quart, but the milk would be given away free to the poor. She later became president of the American Milk Goat Record Association, the influential national organization of goat breeders.[20]

The Oak Forest Tubercular Hospital in Chicago was one of many hospitals and tuberculosis care institutions across the country that began keeping dairy goat herds or purchasing goat's milk for their patients. Near

Philadelphia, the private Radnor-Wayne Sanatorium kept a herd of goats along with dairy cows; "goat's milk being highly recommended in the treatment of consumptive diseases, it was considered highly necessary that such a dairy should be established in connection with the institution[.]" In Texas, the Kerrville Sanitarium near San Antonio purchased dairy goats from Southern California. William Secor, a doctor involved with the Texas project, was a firm believer in the administration of goat's milk, noting floridly that "the time is not too distant when every sanitarium receiving gastro-intestinal and rest cure cases will consider a milk goat herd as a part of its armamentarium." North Dakota's San Haven Sanitorium built its own goat barn and imported fifty Toggenburg goats from Switzerland in order to establish a source of healthy milk for children in its care. The sanitarium at Las Vegas Hot Springs in New Mexico kept a herd of cows as well as an additional forty dairy goats "whose milk [is] served regularly at table, free of charge, to those who desire it."[21]

A tuberculosis patient seeking care in California was more likely to find a health facility that served goat's milk than anywhere else in the nation. California had long been a destination for patients since the time when the disease was known as consumption, when travel, sunlight, and fresh air were considered the ideal therapy for lung problems caused by the disease. As railroads extended their reach to the Western states in the late nineteenth century, railroad companies promoted the West as a health destination in part to benefit their own bottom line. A number of states and territories, including Colorado and New Mexico, were popular destinations for consumptives, but Southern California, with its ubiquitous sunshine and dry air, was a particularly attractive destination for health seekers. The city of Pasadena, California, was founded by a group of consumptive patients from Indiana, who sent scout D. M. Berry to the region in 1873. Pasadena, first known as the "Indiana Colony," along with other cities in Southern California, grew quickly as consumptives flocked to the region.[22]

Among the numerous facilities in California that catered to tuberculosis sufferers, the Martyn Sanitorium in Pasadena acquired goats from prominent Toggenburg breeder Winthrop Howland, whose El Chivar Ranch was based in Redlands, just sixty miles to the east. A number of other institutions marketed the availability of goat's milk, including Casa Desierto Rest Camp in Lancaster, California, north of Los Angeles, which advertised its "pure air, good food, goat's milk, comfortable rooms" in area medical journals. In Long Beach, Arthur E. Pike's Osteopathic Health Resort delineated the available milk types by both species and breed, noting the availability of both Holstein cow's milk and Toggenburg goat's milk to prospective patients in newspapers ads. Even larger medical institutions, like San Francisco General Hospital and Santa Barbara General Hospital, purchased goat milk from local goat dairies for patients in their care.[23]

❖

Goat's milk also became popular for infant feeding. A variety of mammals' milks have been used to feed human infants throughout history. Goats in particular have historically been employed as wet nurses when the animals were more readily available than cows or when infants appeared intolerant of other types of milk. So it was not a conceptual stretch for parents and doctors to feed infants goat's milk, especially as it was believed to be free of if not protective against tuberculosis. Among the advocates for the use of goat's milk for infant feeding was Harvey W. Wiley, the first head of the Food and Drug Administration (FDA), sometimes called the Father of the Food and Drug Act. After his career at the FDA, Wiley assumed a position at *Good Housekeeping* magazine, where he wrote an influential column and contributed to the establishment of the magazine's well-known "seal of approval" for household products and practices. As part of a comprehensive advice essay on infant feeding, Wiley opined:

> It would be advisable, in my opinion, to introduce the cultivation of milk goat in the United States. The milk of the goat does not differ greatly from that of the cow. The goat has this advantage over the cow: it rarely suffers from tuberculosis. Hence one of the most threatening aspects of careless infant feeding would be eliminated if goat's milk were used generally instead of cow's milk. ... Instead of building mills to grind cereals and sprouting barley to form malt to convert these cereals into so-called infant foods, it would be far wiser to introduce a good type of [dairy] goat into every community.[24]

Research institutions began to investigate the suitability of goat's milk for infant feeding. In 1908, the Michael Reese Hospital in Chicago performed experiments on a variety of milks fed to infants in its care. The hospital purchased fifteen common goats and bred them with a purebred Nubian buck, a breed known for its dairying capabilities. Researchers administered a number of types of milk, including mother's milk, milk from a wet nurse, cow's milk of various breeds, and goat's milk, and observed the effects of each on the infants in its care. The goats themselves were kept outdoors on the hospital grounds. During the summer, the babies were also kept outdoors under a big tent, and the goats were milked as soon as the infants cried, getting their "warm milk under the cleanest and most antiseptic conditions, and the youngsters prospered amazingly."

Meanwhile, the emerging field of nutrition science generated more positive information about the value of goat's milk. Some analyses of milk and its various components showed that goat's milk was the most similar milk to human milk, making it preferable for infant feeding. Research conducted at Michael Reese Hospital determined that the fat globules of goat's milk are similar in size to mother's milk (considered a positive characteristic). The hospital also compared goat's milk to various kinds of cow's milk; Jersey cow's milk in particular was deemed unsuitable for babies, as it fat globules were so large that the milk "prevents the stomach from performing [its] right function."[25]

The New York Agricultural Experiment Station in Geneva, New York, acquired a donated herd of dairy goats in 1910. Station researchers partnered with doctors at St. Mary's Infant Asylum and Maternity Hospital in nearby Buffalo to study the suitability of goat's milk for infants at the hospital. Like researchers at Michael Reese Hospital, researchers concluded that for infants who were ill or otherwise failing to thrive, goat's milk was superior because its smaller, more flocculent curds were less irritating to stomachs, allowing the nutrients to be more easily absorbed. The station distributed its excess goat milk for free to individuals in need of milk for infants on the condition that they would report back as to the results. One of those reports read in part: "We certainly appreciate your kindness in furnishing the [goat's] milk . . . I am delighted to say that our baby had gained in one month three and one half pounds in weight, which is quite remarkable for her as for months her weight was the same, and she could not take the cow's milk." Citizen reports were so uniformly positive that extension researchers remarked, "evidence so marked, even if given on an unprofessional basis, is not to be disregarded." As interest in goat's milk grew, its potential public health importance began to attract government attention and resources.[26]

Goats for Uncle Sam

While thousands of migrants traveled West across the continent during the nineteenth century, either to California for the Gold Rush or to Oregon Territory via the Oregon Trail, Kansas native George Fayette Thompson went the opposite direction, traveling east to Washington, DC, where he started his career as a clerk at the United States Bureau of Animal Industry in 1894. The United States Department of Agriculture established the BAI in 1884, charging the subdivision with researching and eradicating diseases of the nation's animals. Over the years of its existence, the bureau released dozens of authoritative educational booklets on a range of subjects, from Circular 68, *Diseases of the Stomach and Bowels of Cattle*, to Bulletin 63, *Foot Rot in Sheep: Its Nature, Cause and Treatment*. The accumulating body of research of the bureau, alongside that of the Department of Agriculture generally, began to generate a knowledge base through which the government educated and informed farmers across the nation. The Hatch Act of 1887 furthered this goal when, among other things, it established state agricultural extension offices in cooperation with state universities across the country. These institutions made significant contributions to the USDA's mission of advancing national agricultural progress.[27]

Although the BAI had been heavily involved in livestock research since its inception, the bureau's researchers were slower to take up the subject of goats. In 1898, the Department of Agriculture issued a goat-related circular titled *Keeping Goats for Profit* by Almont Barnes, a retired Union army colonel working for the agency's Division of Statistics. The publication outlined the ins and outs of raising Angora goats for mohair production, then a growing industry in many Western states.

The Angora goat circular proved to be quite popular, and clerk George Thompson was enlisted to handle the heavy volume of correspondence associated with it. While not a scientist (his position at the BAI was listed as clerk and later editor), George Thompson became the bureau's default goat expert because of his work with the Barnes circular's correspondence. A few years later, Thompson was conscripted to write a more definitive BAI publication on Angora goats, *Information Concerning The Angora Goat*, issued in 1901. Finally, in 1905, dairy goats merited their own BAI bulletin, *Information Concerning the Milch Goats*, also authored by Thompson. Unfortunately, *Information Concerning the Angora Goat* was later judged to have been substantially plagiarized from Texas rancher William L. Black's book, *A New Industry, or, Raising the Angora Goat, and Mohair, for Profit*, published in 1900. After Black mounted a legal challenge, Congress eventually awarded him $5,120 in damages for the plagiarism.[28]

Nevertheless, as momentum and interest in dairy goats and goat's milk grew across the country, the BAI began to take the study of goats more seriously. Daniel E. Salmon, then chief of the bureau, resolved to establish a herd of goats for research purposes. Toward that end, Salmon enlisted George Thompson to travel to Europe to select the best dairy breeds and bring them back to the United States for study. Thompson determined that goats from the island of Malta were the best animal research subjects for the bureau and subsequently traveled there, where he purchased sixty-one female and four male goats to bring back to the United States via ship. On the return trip, however, a number of the ship's crew and passengers fell ill; many of those who became ill tested positive for Malta fever (also known as brucellosis), because they had consumed milk from the goats onboard. Thompson himself died just a few months after his return to the United States. While Thompson's death was attributed to pneumonia, the cause was likely Malta fever contracted by consuming milk from the infected goats.[29]

The burgeoning goat dairy industry lost an influential early champion with the death of George Thompson. Yet neither the loss of Thompson, nor the awkward circumstances surrounding his death, impeded the government's goat dairying and research interest. The USDA went on to construct a goat barn amidst other livestock facilities at its newly established research farm in Beltsville, Maryland. The initial goat herd at Beltsville consisted of forty-four so-called common goats. According to a bureau scientist writing in *The Goat World* magazine, "It was hoped that a strain of goats might be developed by selection, entirely from native stock, which would prove profitable for milk production." After several

Saanen and Toggenburg dairy goats at the USDA Agricultural Research Center in Beltsville, Maryland, 1919. National Photo Company Collection, Library of Congress.

years, researchers found that they were not making satisfactory progress in breeding a high-producing line of dairy goats, so the bureau acquired European purebred Saanen and Toggenburg bucks for their productive dairy genetics. Milk from the government dairy research operation was sent to nearby Georgetown University, where it was used for research on the topic of infant feeding. By 1925, the Beltsville herd had grown to fifty goats. The Bureau of Animal Industry's acquisition of goats represented a significant milestone for the cause and credibility of the goat dairy industry in the United States.[30]

❖

The nineteenth and early twentieth centuries brought fundamental changes to human understanding of the nature of human illness and disease. Robert Koch's discovery of the tuberculosis bacterium had widespread implications not only for human health, but for the nation's milk supply as well. Once scientists began to understand that cattle could pass the tuberculosis through their meat and milk, the American public began to turn its attention toward another milk source: the goat. The emerging positive perception of goat's milk had a secondary effect of changing the largely negative reputation of the animals.

One additional note: in recent years, scientists have definitively answered the question of whether goats are immune from tuberculosis. The answer is no. Goats can carry the *M. bovis* strain as well as another tuberculosis strain known as *M. microti*. Infected goats can pass the disease to other animals as well as humans. Infected goats can also pass tuberculosis through their milk as well. Modern state and federal laws require that dairy animals, including goats, be tested for tuberculosis and brucellosis in order for their milk to be sold commercially. Scientists have also identified a goat-specific strain of tuberculosis known as *mycobacterium caprae*, which has been isolated primarily in Europe and is particularly prevalent in in Spain. According to one study, "Compelling evidence indicates that *M. Caprae* poses a serious health risk not only for goats, but also for other domestic and wild animal species and humans."[31]

CHAPTER FOUR
The Goat's Milk Business

I believe that within ten years the milk goat will be to the nation what the Ford car is to the motor industry today.
    Charles A. Stevens, prominent Chicago businessman, 1923

The burgeoning goat's milk movement of the early twentieth century found an early champion in Lelia Foster Roby. Referred to at the time as Mrs. Edward Roby and wife of a prominent Chicago lawyer, she possessed her own impeccable pedigree. A Boston native, she could trace her family lineage back to the *Mayflower*. In an early twentieth-century book about extraordinary women in American history, it was said that "America has hardly produced a woman of better courage and patriotism" than Mrs. Roby. During her life she was an active member of numerous philanthropic organizations, including the Daughters of the American Revolution. She founded the Ladies of the Grand Army of the Republic, an organization devoted to Union Civil War veterans. Mrs. Roby represented the Chicago Board of Education before the Illinois state legislature and once lobbied Secretary of State Richard Olney on behalf of another organization, the United States Daughters of 1812. As if all of that wasn't enough, a double fringed crimson petunia with white edges was named after her. Lelia Roby was one of a new generation of wealthy, well-connected Gilded Age philanthropists.[1]

Chicago philanthropist Lelia Roby. Photo from Francis Willard and Mary A. Livermore, eds., *A Woman of the Century: Leading American Women From All Walks of Life* (Charles Wells Moulton, 1893).

Among her myriad charitable endeavors, Mrs. Roby became interested in dairy goats. According to contemporary reports, "[i]t is Mrs. Roby's purpose to sell goats to households that the children will be better nourished and relieved of the danger of contaminated milk." From Roby's perspective, the broader social problem was the public's access to pure, fresh milk. But Roby did not decide to sell or distribute milk; instead she conceived an ambitious plan to develop an ideal American goat breed, an animal of hardy constitution that was also a good milker and could be bred and distributed to the poor (notably, she does not appear to have expressed interest in drinking goat's milk herself, at least not publicly). Toward this end, Mrs. Roby started a farm where she kept goats and began the process of selecting the best milkers among them. From these she hoped to establish a breed of goat that might subsequently become the foundation of a domestic dairy goat industry. Roby attracted some attention for her efforts; one of her most productive goats, named Watita, was featured in the influential Bureau of Animal Industry (BAI) publication *Information Concerning the Milch Goats*. Author George Thompson (see chapter 3) was supportive of Roby's efforts, remarking that Watita's "conformation and record show her to be a very desirable animal as one of the mothers of the American milch goat."[2]

While Mrs. Roby's intentions were philanthropic, her efforts diverged somewhat from prevailing animal husbandry practices. We often think of domestic animals in terms of a breed identity—a Labrador Retriever dog, a Jersey cow, a Rhode Island Red chicken. While certain animals have been individually selected for desirable characteristics for centuries, the idea of specific animal *breeds* is a more modern development. The roots of the practice of selective breeding now commonplace in the Western world are generally traced to eighteenth-century sheep and cattle rancher Robert Bakewell of Leicestershire, England. Bakewell was a careful observer who noted desirable traits in his livestock, such as wool production in sheep or body size for meat purposes in cattle, and then bred his animals to perpetuate those specific traits. Eventually, successive generations of progeny displayed the preferred attributes; Bakewell's cattle were heavier and his sheep produced more wool than those of his neighbors. Bakewell's "improved" animals, which represented a distillation of desirable inherited characteristics, became known as a breed of animal—not just a sheep, but a Leicester sheep. When a farmer possessed a specific breed of animal, that farmer had in effect bypassed decades of individual selection and arrived, more or less instantly, at the best possible expression of a given desirable trait. Gregor Mendel's nineteenth-century experiments with pea plants led to a more detailed understanding of the mechanisms of genetic inheritance, allowing breeders to predict the patterns of livestock breeding with even more precision.

While Europeans and Americans had experimented with breeding all manner of livestock for over a century after Bakewell's efforts, it was not until the nineteenth century that American livestock raisers began to seriously ponder the question of goat breeds. Until that point, every goat in the United States was considered to be of more or less undetermined ancestry, descended from those transported to the Western Hemisphere from European countries like Spain, England, France, and the Netherlands starting in the sixteenth century. During the early days of American colonization there was little concern for goat breeding, as the animals carried such negative associations (see chapter 1). As George Thompson of the BAI observed in *Information Concerning the Milch Goat*, "There appears nowhere to have been an effort to keep the blood of [goats in this country] pure, and any virtue that there might have been in any of the breeds imported was soon dissipated."[3]

But as science and the general public began to realize that cow's milk was potentially contaminated and even dangerous, concerned citizens sought out goats and their milk as what they hoped was a safe, healthy alternative. At the same time, in an era that valued purebred livestock, simply rounding up American goats of undetermined origin and calling them dairy goats or spending years carefully selecting them for milk production (the path more or less outlined by Mrs. Roby) would not be sufficient. Instead, livestock breeders began importing European dairy goats into the United States. William A. Shafor, a second-generation livestock breeder from Ohio, was active in several organizations, including the Polled Hereford Association (named after the cattle breed) and Oxford Down Sheep Association. Shafor imported what are believed to be the first European dairy goats into the United States from the Toggenburg Valley in Switzerland, by way of Britain, in 1893. While Shafor does not appear to have been interested in breeding and keeping goats himself, his importation efforts, said to have been made on behalf of a friend, were the earliest of what eventually grew into a significant influx of imported European dairy goats. For the next several decades, a variety of goat breeds, including Saanen, French Alpine, and Schwarzenburg, from countries such as Switzerland, France, and Germany were imported into the United States.[4]

The influx of European dairy goat breeds elevated the humble goat to a new level of legitimacy, on a level with prized breeds of cattle and sheep. European breeds facilitated the transformation of the public's perception of goats beyond centuries-old stereotypes. Contemporary descriptions reflect this distinct shift in the perception of goats by both the media and the public:

> The plutocratic milk goat is the exact antithesis of the ordinary tin can alley goat. The blue blooded goats are disciples of sanitation. They will eat nothing but clean and pure food. They require sanitary stables and yards. Personally, they keep their shaggy overcoats as spotless and immaculate as those of the most particular "tabbies" of the feline world.

European goats elevated the very concept of the animal to a hardy, utilitarian, and even regal stature and status. These were not ordinary goats; these were *dairy goats*.[5]

❖

World's fairs were once extravagant global events. Part grand exhibition, part international self-promotion, the celebrations were held periodically in cities across the world. During their peak of prestige and influence in the late nineteenth and early twentieth centuries, the fairs served as cultural platforms for nations to display advances in technology, design, and culture. We have world's fairs to thank for the construction of the Eiffel Tower and the debut of the Ferris wheel. Alexander Graham Bell demonstrated his revolutionary new invention, the telephone, at the Philadelphia World's Fair in 1876. In addition to displays of the latest inventions, world's fairs also featured agricultural pavilions where participating nations displayed the latest achievements in farm products, agricultural machinery, livestock breeding, and horticulture. And so it was with a sense of progress and patriotism that the growing community of United States goat fanciers targeted the St. Louis World's Fair of 1904 (also known as the Louisiana Purchase Exposition) as their moment to promote the advent of the nation's dairy goat industry.

But it would not do to present a group of unorganized goat fanciers on the world stage. By some accounts, the World's Fair organizers would not allow unregistered animals to be shown. And so in 1903, George Thompson of the BAI contacted David Tompkins, a goat enthusiast in New Jersey who had previously donated goats to the bureau. Thompson asked Tompkins to become president of a proposed association of goat breeders "until it could be successfully organized." Then, just days before the St. Louis World's Fair, Mrs. Roby, George Thompson, and William A.

Shafor, along with several other goat enthusiasts (Tompkins could not make the meeting), met in the Livestock Congress Hall on the grounds of the St. Louis World's Fair and convened the first meeting of what they called the American Milk Goat Record Association (AMGRA). Edmund P. Cohill, a Maryland farmer whose son William was a goat entrepreneur, became president, Mrs. Roby served as vice president, and William A. Shafor as secretary. George Thompson was officially admitted as a member of the organization "because of his services as a writer on the subject [of goats] . . . and as an official of the Bureau of Animal Industry." Decades later, the group changed its name to the American Dairy Goat Association (ADGA), and the organization continues to thrive in the twenty-first century.[6]

Despite the behind-the-scenes maneuvering among goat enthusiasts in the weeks leading up to the St. Louis World's Fair, the dairy goat show proved to be anticlimactic. In the end, only four imported Toggenburg goats were displayed. They were owned by William J. Cohill, a fifteen-year-old from Maryland whose father had become the president of AMGRA days earlier. The younger Cohill was an enthusiastic goat breeder who started out in the goat business after convincing his father to acquire a few Angora goats to help clear the family's 1,200 acres of land. Soon the family herd expanded to several hundred Angora goats, which they used for breeding and brush control, as well as for mohair production. Cohill eventually turned toward goat dairying and purchased his own imported dairy goats, which he exhibited widely. "There is a little fortune awaiting you in the goat business," Cohill wrote in *The American Boy* magazine.[7]

A growing interest in dairy goats starting in the late nineteenth century led American goat enthusiasts to import European goats into the United States. Not only did European goats carry a sense of Old World legitimacy, but they also contributed dairy specific characteristics that breeders hoped would translate into productivity and profitability. The subsequent founding of AMGRA, an organization devoted specifically to goats and goat breeding, marked a significant step toward the professionalization and legitimization of the new business of goat keeping in the United States. The organization provided a voice and a platform from which goat breeders developed an organized profession, traded ideas and resources, and led to the development of a goat dairying community. The stage was set for the growth of the goat dairy industry in the United States.

The Wise Man's Cow

One of the effects of society's growing attention to the powers of goat's milk is that prevailing negative attitudes about goats began to shift dramatically. Newspapers from across the country suddenly trumpeted positive and uplifting news about goats. Headlines that would have been unimaginable a decade earlier gushed goat positivity without a trace of shame: "Goat No Longer a Joke," announced *The Boston Globe*. The *Fort Wayne Weekly Sentinel* of Indiana advised readers to "Buy a Goat and Eliminate the Doctor Bills." At the same time, a variety of periodicals directed at goat enthusiasts appeared, including *The American Standard Milch Goat Keeper* from Massachusetts, *The Goat World* from Los Angeles, and *The Milk Goat Journal* from Nebraska. These and other publications devoted themselves to the new trending subject of national interest: information and education about dairy goats. The widespread availability of news and information about goats contributed to the development of communities of goat breeders, who in turn began to form regional goat societies and breed networks across the country.[8]

While goat dairy promoters aspired to solve the entrenched milk contamination problems of the day, the movement also dovetailed neatly with another social movement—the back-to-the-land movement. An increasing number of Americans were dissatisfied with the rapid industrialization of the country during the late nineteenth and early twentieth century, a process they believed had moved the country too far away from its agrarian roots. Decades of economic swings, including the depression of 1893, bank panics, and successive waves of widespread unemployment, created a sense among some that the route to human happiness and prosperity was not, in fact, to strive for industrialist riches, but rather to get back to the land, buy a few acres, grow your own food, and take control of your own destiny. Books like Bolton Hall's *Three Acres and Liberty* and its successors urged citizens to free themselves from the chains imposed upon them by industrial capitalism.

W. Sheldon Bull appreciated the nexus of dairy goats and a life of self-sufficiency, at least milk self-sufficiency. In his article, "One Family's Solution of the Milk Problem," Bull, who lived with his family in Buffalo, New York, outlined how he had established his own independent healthy milk supply for his family by purchasing goats. "The only way to become independent of the milkman and his blue milk would be to install a small dairy on our premises . . . we decided to establish a small goat dairy in our backyard." Just as Bolton Hall had advocated the use of vacant lots in cities to grow food, Bull took the concept one step further by recommending that his readers turn their backyards into food sources. In addition to supplying the family with milk, the goats allowed the family an additional income stream. "What milk we do not need for our own household," Bull wrote, "we find no trouble in selling as a nutrient for infants and invalids as supplied by their physicians." Bull's vocal advocacy for dairy goat keeping in urban backyards marks a distinct shift from the decades of efforts to rid urban areas of goats and all the social ills they had come to represent. Goats were clearly beginning to move beyond their tarnished reputation as the mascot of the poor and into a new role: health-giving saviors, not to mention liberators of families across the nation. W. Sheldon Bull later published a book expanding on his goat keeping experiences titled *Money in Goats* in 1911, which continued his enthusiastic goat advocacy while providing advice about goat care.[9]

As interest in goats and their milk grew a new category of small businesses emerged: the goat dairy business. Across the country, goat dairies began to dot the landscape. Among the nation's early goat dairy entrepreneurs was Melvin (M. P.) Eggers, who began his goat dairying career in Woodinville, Washington, outside of Seattle. Eggers began selling goat's milk in Seattle in 1921 under the name Encaria Goat Farm. Later Eggers branched out into making cheese; he sold two types of goat's milk cheese, "brown" (probably resembling gjetost, a Norwegian cheese made from caramelized whey) and "white," through ads in *The Goat World* magazine. Over many years, Eggers advertised his goats and cheese prolifically in *The Goat World* and a number of other goat, agriculture, and livestock publications; his ongoing visibility in these publications are a testament to his tenacity, if not his success, in the goat business.

Many goat dairy businesses incorporated terms like European or Swiss into their names or their product as a way to invoke the aura of purity conferred by European goats. Perhaps the champion in this regard was Southern California's San-a-Tog Dairy, which cleverly combined breed names Saanen and Toggenburg, while also managing to evoke the word "sanitary." Another goat entrepreneur, Frank Habig, founded Swiss Goat Dairy in Indianapolis in the mid-1920s. The Habig family first acquired goats when daughter Ruth was sickly and unable to gain weight as an infant. Ruth's health was restored with goat's milk, and the family soon found themselves in the dairy goat business. Business was so good that in the 1940s that the Swiss Goat Dairy debuted a new building complete with a "dairy bar where goat milk can be bought by the drink." The company grew into an agritourism destination where customers could buy and drink milk, as well as observe and interact with the goats, which were milked on-site. Like Melvin Eggers in Washington, Habig heavily promoted the health-giving properties of goat milk.[10]

In New England, a number of dairies, such as Lone Oak Goat Farm, which advertised itself as a "Swiss Goat Dairy" in Delaware, and Waltham Goat Dairy in Waltham, Massachusetts, just outside of Boston, provided milk to area residents. Denver, Colorado, could boast of at least two goat dairies by the mid-1920s, the Denver Goat Dairy and Crampton Goat Dairy. In the Chicago area, former policeman William A. Haedtler attributed the success of his Haedtler Goat Milk Dairy Farm to the use of "scientific methods and sanitation." In the Atlanta area, Westwyndes Goat Dairy in Chamblee, Georgia, made a big splash with its herd of Nubian goats and on the farm milk sales. Another Atlanta-area goat dairy, Pitts Goat Dairy, was started by Mr. J. O. Pitts because of his stomach trouble. He purchased one goat, and his health improved from the milk. Afterward, he purchased several more, and soon was selling goat's milk all over the city.[11]

California emerged as one of the centers of the commercial goat industry in the United States. The state was often called "the nursery" of the goat dairy industry, and California's dairy goat community made up the largest group of members of the American Dairy Goat Record Association in its early years. In part this was because the climate, especially in more arid Southern California, was well suited to goat rearing. In addition, the Southern California climate had long attracted health seekers, especially tuberculosis patients. For several years, the California Milch Goat Association, founded in 1916, even sponsored a float in the Pasadena Rose Parade. While the Rose Parade was not then the internationally promoted and televised event that it is currently, the goat group clearly hoped to make a promotional splash, at least within the local community. In 1916, the group's first float, complete with live goats, was awarded a silver cup for first prize in its division. Their 1917 entry proclaimed the goat to be "The Little Friend of All the World."[12]

One of the leading figures in the Southern California goat dairying community was Florian "F. T." Heintz. Heintz had moved to Baldwin Park, California, from Texas and became a real estate developer and goat enthusiast. He started *The Goat World* magazine in 1916, which became one of the leading periodicals of the day covering goats, goat keeping, and goat news. But Heintz's enthusiasm for goats was not just in a journalistic capacity; he also sought to develop a large tract of land in Baldwin Park, which he called Vineland Acres. He appears to have envisioned a utopian-style collective colony, advertised in *The Goat World* as "The Largest Goat Colony in the World," where like-minded goat breeders could live near one another and work cooperatively. Half-acre lots were available for $500–$900 and came with the additional advantage of being adjacent to the Baldwin Park Goat Cheese Factory, which Heintz had established on the property.

California's livestock industry, including its goat enthusiasts, suffered a significant setback in 1924 when an outbreak of hoof-and-mouth disease spread throughout the state. State health officials instituted a strict quarantine, ending California goat breeders ability to sell goats and breeding stock, a significant source of their income. News of the outbreak effectively ended Heintz's plans for selling lots in his goat utopia. Heintz eventually resigned as editor of *The Goat World*, and later ads for Vineland lots advertised their attractiveness for chickens rather than goats. Heintz Street in Baldwin Park is all that's left of that Los Angeles suburb's goat dairying history.

The emerging goat dairy industry attracted many entrepreneurs with outsized aspirations. For some, the economic possibilities seemed attractively open-ended. In the Pacific Northwest, for example, physicians J. W. Morrow of Portland, Oregon, and Hy Parker of Arizona conceived of a plan to establish a goat dairy on Cypress Island, a mostly uninhabited island in the San Juan Islands chain off the Washington coast north of Seattle. A local newspaper proclaimed enthusiastically, "Islands to be Nation's Goat Center." The pair's objectives changed over time: the men appear to have first intended to start a goat dairy that would supply milk to the Seattle area, then pivoted to the idea of developing a sanitarium-like resort with goats, and later hoped to create a goat's milk cheese factory. The small city of Anacortes, Washington, even considered establishing a tuberculosis sanitarium on Cypress Island because of the potential availability of goat's milk there. The men purchased several hundred Saanen, Toggenburg, and Nubian goats in Arizona and brought the animals by train to Bellingham, Washington, where they were then transported by boat to the island. Drs. Morrow and Parker hoped to accumulate a herd of some five or six thousand goats simply through the natural multiplication of this initial stock. Perhaps the physicians were not prepared for the practical aspects of a goat dairy operation. The business changed hands several times amidst reports that hundreds of goats had been left to run wild on the island. At one point, one of the farm's herdsman was arrested for smuggling forty cases of liquor from Canada to the United States. Running bootleg liquor from Canada through the San Juan Islands to the mainland during Prohibition was a common practice of the period, and also apparently an attractive sidelight for isolated goat herders seeking supplementary income.[13]

Meanwhile, serial entrepreneur A. B. Hulit convinced a group of Midwest physicians to buy stock in the National Goat Dairy Company in 1904. Hulit, an Angora goat breeder, had ambitious plans to establish a large goat dairy near St. Louis, which would supply the city with goat's milk. According to reports, Hulit was set to travel to Europe to procure five hundred purebred goats for the group, which were to be exhibited at the St. Louis World's Fair, though nothing ever came of the venture. A similar effort mounted by a group of partners in Petaluma, California, who advertised their efforts to establish the California Swiss Goat Dairy, the "Coming Industry of California," in 1906. The ambitious group promised to create a facility that would supply goat's milk to local hospitals, as well a cheese factory, though nothing ever came of that venture either. These and other failed efforts of the period demonstrate that despite the public's growing enthusiasm for goat's milk, the practical aspects of starting and maintaining a goat farm and dairy were considerable and not always appreciated by those without farming experience.

That being said, for the very wealthy, the financial and practical challenges of starting a goat dairy were easily overcome. Gentleman farmer Charles A. Stevens, owner of the Chicago department store chain of the same name, became interested in goats when his granddaughter fell ill and a physician prescribed goat's milk. Stevens said that when he had attempted to buy goat's milk, he found none for sale in the Chicago area. Stevens happened to own a country estate, which called Agawam, in Delevan, Wisconsin, about one hundred miles north of Chicago. He subsequently acquired a few dairy goats, and eventually became enthusiastic enough about the industry's potential to become involved with the Cook County Goat Commission alongside fellow Chicago area goat enthusiast Mrs. James Patten.[14]

As Stevens became increasingly involved in the business of goat breeding, many in the region began to seek out both milk and goats from him, and the astute businessman saw an opportunity. Stevens devised a plan to acquire a number of prized Toggenburg goats, the most popular goat breed of the early twentieth century, in Southern California, transport them by train to his Wisconsin farm, set up an auction, and have his wealthy friends bid on the herd. Stevens knew marketing, and so naturally the entire process was highly publicized. The *Los Angeles Times* announced the event: "Southlands Goats to Stock Chicago Millionaires' Dairies." The article described in great detail the sendoff reception for Stevens's goats in Los Angeles, which was attended by Southern California goat luminaries F. T. Heintz; Winthrop and Martha Howland of El Chivar Ranch in Redlands;

Howland's sister-in-law Jane Storey White, also a prominent Southern California goat breeder; as well as Violet Kirby and Rose Saunders, owners of Canyon Goat Ranch in Redlands. The rail cars carrying the goats were reported to have been outfitted luxuriously, with an accompanying herdsman and assistants, a veterinary area, and plenty of additional room for any kids born along the way. Stevens even designed the rail journey to make a big splash along the route eastward, as the goats were to be milked along the way and their milk distributed at train stops.

Stevens acquired seventy purebred Toggenburg goats in California. As promised, he arranged a high-profile auction of the goats at Agawam, going so far as to charter a special train to bring bidders from Chicago to his ranch, outfitted with two dining cars and a barber shop, among other amenities. The bidders, all Chicago area Gold Coast millionaires, included William Wrigley Jr., Mrs. James Patten, and A. Watson Armour of the Armour meatpacking dynasty. Armour had his own farm estate called Elawa Farm in Lake Forest, north of Chicago where, like Stevens, he kept a variety of livestock including dairy goats. *The Goat World* heralded the Stevens auction as transformational, "elevat[ing] the humble goat . . . to the greenest pastures of rich men's country estates." The event, as well as the widespread national publicity generated by Stevens, was certainly a boon for the nascent industry.[15]

❖

Until the advent of refrigeration, storing and transporting fresh milk, a highly perishable product, was not a simple task. While cheese and other types of fermented dairy products served as ways of preserving and storing milk for centuries, inventors of the industrial age harnessed the powers of heat and steel to create a new milk storage technology: condensation. Gail Borden patented the first milk condensation apparatus in 1856, a moment that marked the inauguration of the canned milk industry. Canned milk revolutionized milk production and distribution since the canned product was easy to transport and had a long shelf life.

Around the same time, Swiss scientist John B. Meyenberg independently developed his own condensation process. Meyenberg's process differed from Borden's in that his formula did not contain sugar. Although the products are similar, the unsweetened version of condensed milk is typically called evaporated milk. When the Swiss company Meyenberg worked for ignored his invention, he moved to the United States and patented his technology. Meyenberg then started the Helvetia Milk

Condensing Company in Illinois, and he eventually sold that company to the PET milk company (now owned by General Mills). Meyenberg went on to work as a consultant for a number of companies, including the Pacific Coast Condensed Milk Company in Kent, Washington, which became the Carnation Milk Company (now owned by Nestlé). Eventually Meyenberg and his family relocated to Monterey County, California, and started their own condensed milk company in 1906.[16]

The rapid growth of the cow's milk condensation industry across the United States set the stage for a parallel goat's milk condensation enterprise. One of California's early twentieth-century goat's milk entrepreneurs was Chris (C. H.) Widemann of Gonzales, California. He was the son of Alfred Widemann, who started the regional chain of A. Widemann department stores. C. H. Widemann was often described as "restless" in the contemporary press, and the reasons are apparent: he had his hands in dozens of businesses throughout his career, including an oil company and numerous land deals across California. In 1916, Widemann turned his attention to goats, partnering with Walter Meyenberg (one of the three sons of John B. Meyenberg) and local businessman George Miller in founding the Widemann Goat's Milk Company. Widemann aspired to produce a line of evaporated goat's milk in cans.

C. H. Widemann understood that one of the biggest challenges of the evaporated goat's milk business would be obtaining a reliable milk supply,

GOAT MILK–PURE
A Baby Food without a substitute
Universally used in Europe
Highly Recommended diet for Stomach Trouble and Tuberculars.

World's only producers of Evaporated, Unsweetened, Pure Goat Milk

Two dozen Cans to Case. 20c Per Can. Ask your Druggist.

An Unexcelled Baby Food

Dairy and Condensory at King City, California
WIDEMANN GOAT MIK CO. Gen. Office, San Francisco
212-14 Physicians' Bldg.

The Widemann Company often emphasized in ads that its evaporated goat's milk was especially suitable for tuberculosis patients. *Angora and Milch Goat Bulletin*, October 1916.

and so he developed a large goat farming operation. Descriptions of the period paint an impressive picture. He secured a lease on a 3,000-acre tract in King City, California, owned by Loren Coburn, who had made his way to California from Vermont during the Gold Rush years. Coburn made his fortune with several businesses in the Bay Area and went on to deploy his considerable wealth in the land speculation business, purchasing thousands of acres of land in San Mateo and Monterey Counties. At its peak the Widemann ranch was said to have amassed thousands of goats—reports range from 3,000–6,000, though those numbers are likely exaggerated—including goats imported from the Southwest and Mexico. Widemann enlisted prominent California goat expert Irmagarde Richards to manage the substantial operation, which was reported to have produced as many as 1,500 cans of evaporated goat's milk per day.[17]

One of the Widemann Company's strengths was its marketing. Among its innovative sales strategies was to sell its evaporated goat's milk product through drug stores, which lent an aura of healthfulness and medical approval to the brand. While the company was criticized for this move, it was an ingenious way to entice customers by creating an impression that the product was somehow better than everyday goat's milk. In an era when the threat of potentially contaminated milk was at the forefront of everyone's mind, the strategy suggested that the Widemann product was safe and superior. Widemann also marketed its product to San Francisco's growing Italian community, a community that was already receptive to goat's milk, by placing ads in the locally circulated *L'Italia* newspaper for its "Latte Puro di Capra." The ads listed the local drug stores where the product could be purchased.

The business of goat farming, not to mention production and distribution of evaporated goat's milk, proved to be a roller coaster, and the Widemann operation weathered highs and lows over the course of its relatively short existence. In 1917, the company announced that it had been awarded a government contract to produce 15,000 cases of evaporated goat's milk for the war effort. Then Widemann relocated the entire operation, goats and all, from Monterey County north to Pescadero in San Mateo County, presumably to be closer to the larger customer base in San Francisco. Business partner Walter Meyenberg took over management of the operation. More dramatically, C. H. Widemann became immersed in scandal in 1919 when Mrs. Coburn, wife of Widemann's landlord Loren Coburn, who had passed away a year earlier, was found dead. The scandal over her murder and the disposition of the substantial estate raged on for years. Widemann happened to be one of two beneficiaries of her substantial will. At one point the police even sought to arrest C. H. Widemann for the murder of Mrs. Coburn, though a judge ultimately refused to issue a warrant. A fire later destroyed the Widemann condensation facility and two large warehouses; by the late 1920s, Widemann's condensed goat's milk faded from the market. The mystery of who killed Mrs. Coburn was never solved.[18]

Despite the drama, all was not lost for canned goat's milk production in California. After the demise of the Widemann Company, the Meyenberg family picked up the goat's milk condensation business and within a few years started producing evaporated goat's milk under the Alpure and Lullaby brands. According to some contemporary reports John B. Meyenberg's son, John P. Meyenberg, was sensitive to cow's milk, a fact that had originally prompted the family's interest in goat's milk. The Meyenbergs subsequently established a goat's milk evaporation plant in Paso Robles, California, and production was later relocated east to Ripon, California, along with the company's cow's milk evaporation operations. Over the course of the twentieth century, Meyenberg would become one of the most recognized goat's milk brands in the United States.[19]

Women of the Goat Dairy Industry

One of the most prominent figures of the early twentieth-century goat dairying community was Irmagarde Richards. Richards grew up in Spokane, Washington, and graduated from Stanford University in 1902, where she played on the women's basketball team. Richards went on to teach Greek and Latin at Mills College in nearby Oakland, California, and within a few years she had established herself as one of the leading authorities on dairy goats in the country.

Richards came of age during a period when women were beginning to break away from social mores that relegated their roles to a traditional sphere—the home and its associated domestic duties. With the advent of temperance movement, the fight for women's suffrage, and other social causes, women increasingly became involved in issues outside the home in order to address a broader range of social ills, including those which prevented women from having a voice.

The goat dairy industry was attractive to women for several reasons. In the eyes of society, their participation seemed a natural fit in certain professional areas, such as those involving human health, especially infant nutrition. As one advocate noted, "[W]ide demand for pure non-infected milk has resulted in the opening up of a new occupation for women." Much in the same way that women were beginning to join the medical profession or the new and emerging field of home economics, goat raising became an increasingly acceptable avenue for women who aspired to something beyond traditional social norms.[20]

For some women, the early phase of the back-to-the-land movement, with its message of freedom from traditional economic and social constraints, spoke directly to their desire to move away from traditional gender norms. A generation of female farmers, some single women, some working together, found their way to economic independence by raising goats. A contemporary article titled "California Women Goat Ranchers," lists eighteen women in the goat business in that state in 1920.

Irmagarde Richards said she first became interested in goats while traveling in Italy. The passing interest turned to reality in 1915 when Richards and former student turned business and life partner Morris Wagner started Las Cabritas Farm in Montara, California, near Half Moon Bay, where Wagner had grown up. Wagner and Richards built a substantial herd of over two hundred Toggenburg goats and developed a profitable dairy business selling their milk to San Francisco General Hospital and to the Widemann evaporated milk operation in Central California. Richards cemented her reputation in the dairy goat world with her book *Modern Milk Goats*, published in 1921. The book served both a modern guide to goat keeping, as well as a voice for the industry, and became the era's go-to guide for aspiring goat keepers who yearned for practical information about goats and their care.

LAS CABRITAS TOGGENBURGS
A PURE-BRED HERD, OFFERING

The best breed type, attained by years of careful breeding and selection, robust constitution, maintained by the freedom of open range for does and bucks. High average milk production, necessitated by a dairy contract that demands a large yield of milk every day of the year.

LAS CABRITAS GOATS FURNISH THE MILK USED IN THE TUBERCULOSIS HOSPITAL OF THE CITY OF SAN FRANCISCO.

We offer from the increase of this herd the finest breeding stock, at prices practical for dairymen and householders.

Yearling bucks and buck kids from Advanced Registry does, mature does and doe kids. Drop us a postal asking for our sale catalogue.

Irmagarde Richards, Morris Wagner, Montara, San Mateo County, California.

Irmagarde Richards and Morris Wagner raised Toggenburg goat at Las Cabritas Farm in Montara, California. Angora and *Goat Milk Journal*, October 1921.

While goat raising was gaining broad general acceptance during this period, prevailing attitudes about women and farming were patronizing at best. The idea of a woman farmer was generally considered an oxymoron, as farm labor was thought of as too difficult for the average woman. Many thought of goat raising as acceptable for women simply because the animals were smaller and easier to manage. In *Modern Milk Goats*, however, Richards does not mince words about the challenges of the goat-raising business: "Only a woman of steady, dependable health and actual muscular strength and endurance can hope to carry on the goat business on a scale large enough to produce a profitable livelihood," she wrote, adding "[a goat woman] must also be a good sales woman, with a gift for writing advertisements and good selling letters, a capable judge of human nature—in short, she must be a successful merchant as well as a successful breeder." Operating a goat dairy was not for the faint of heart—in other words, goat farming was ideal for a new generation of strong, independent women.[21]

Richards's goat expertise became particularly useful during one particularly high-profile assignment. When the Argentinian consul contacted the University of California for help in purchasing purebred dairy goats to be exported to Argentina in 1923, the school tapped Richards for help, as she was "the person who knows more about goats than anyone else in California." Richards subsequently became Argentina's designated goat broker and scoured the state of California for high-quality dairy stock. She eventually purchased nineteen Saanen and twenty-four Toggenburg goats from the state's breeders and outfitted their quarters on a cargo ship that would transport the goats, and Richards, on the six-week journey to Argentina. The traveling quarters were said to have been well appointed, with individual goat stalls, plenty of hay, and round the clock care provided by Richards and attendants. Despite the fact that Richards had been recommended as the most competent person for the job, considerable diplomatic interventions had to be made in order for the Argentinians to accept a female sales agent. According to reports, President Warren G. Harding's office had to intervene in order to ensure that Richards could make the journey to South America in a cabin instead of riding in the ship's steerage area alongside the goats.[22]

Modern Milk Goats by Irmagarde Richards, published in 1921 by J. B. Lippincott, was one of the earliest authoritative guides to goat rearing and dairying. Collection of the author.

Richards went on to write numerous articles on goats and their care for a succession of goat and agricultural publications, became president of the Central California Goat Breeders Association, judged goats at shows, and spoke at agricultural conferences for a number of years. An educator at her core, later in her life Richards wrote a series of books about the history of California intended for school-age students.[23]

While Richards was the most prominent of the women in the goat industry of the early twentieth century, there were many others. A good number of the era's women goat breeders were concentrated in Southern California. When prominent goat breeder Winthrop Howland, who suffered from tuberculosis, became ill, his wife Martha took over the management of their farm. Goat raising ran in the family; Martha's sister Jane Storey White was also a goat breeder and owned Fair Hope Ranch in La Crescenta, California. Violet Kirby and Rose Saunders, college friends from England, set up Canyon Ranch goat farm on eighty acres in Redlands, near Howland's El Chivar Ranch. Both women had initially traveled to California because of their health. The pair first raised chickens but quickly turned to goats, which they found "far more intelligent." Mrs. Howland tutored the pair in the care and maintenance of the animals. Canyon Ranch became well-known throughout the goat world for its pedigreed Toggenburgs, which the women sold all over the country and abroad through the 1940s. For many years, Saunders also led the Southern California Goat Association. The activities of Irmagarde Richards and the many women goat entrepreneurs across the country provided an essential contribution to the growth and development of the dairy goat industry in the United States.[24]

Cheese From Goat's Milk

As dairy goats became increasingly popular, goat keepers across the country began to seek ways to profit from the milk of their animals. Those not located near large population centers or who were otherwise unable to develop a fluid milk business started to experiment with making cheese. Goat-centric publications like *The Goat World* promoted cheesemaking to its readers, publishing recipes for making cheese and even suggesting that readers could make more money selling goat's milk cheeses than they would with cow's milk cheese. Numerous individuals took out ads in newspapers, as well as dairy goat publications, offering their home-produced goat cheese products by mail order.

Humboldt County, California, is known as the home of popular cheesemaker Cypress Grove Cheese Company, which was founded in 1983, but that company was not the first commercial goat cheese factory in the county. During the early 1920s, a group of goat owners in Ettersburg, south of Eureka, led by Toggenburg breeders James E. French and Alfred E. Pixton, formed the North Counties Milk Goat Association, likely the first formally organized goat's milk cooperative in the country. Because the area was distant from big population centers where demand for goat's milk was highest (Ettersburg is 225 miles north of San Francisco on today's paved roads), the group decided to produce cheese instead. The Ettersburg factory, which the *Humboldt Times* proudly proclaimed the "first pure goat cheese plant in the United States," had serious ambitions; the group hired a Swiss cheesemaker and produced a cheese it called Eurisco, said to contain "all of the rich nutritive qualities of goats milk in concentrated form."[25]

Even as goat owners like those in Ettersburg were turning their attention to making cheese, immigrants across the country were busily forging their own commercial goat cheese industry. Italian immigrant families in the Paterson, New Jersey, area herded goats and made cheese during the first decade of the twentieth century; contemporary reports mention as many as 1,000 goats. The region's large Italian immigrant population provided a significant market for the cheese and dairy products produced there. Paterson's Scheps family (shortened from the Italian Schepis) went on to operate a large commercial ricotta and mozzarella manufacturing company through the 1980s. In New Hampshire, Nicholas J. Nassaikas, known to some as "the Goat King of New England," kept a herd of 600 goats on his 850-acre ranch near Manchester. Nassaikas said he owned a 10,000-acre farm in Greece where he kept goats and made cheese, which was imported to the United States. When the onset of World War I got in the way of his imports, Nassaikas began to make cheese and butter from the milk of his extensive New Hampshire herd.[26]

During the late nineteenth and early twentieth century, mining, railroad, and ranching companies recruited immigrants then arriving by the thousands at East Coast ports to travel to jobs in Western states and territories. Italians working at Utah's mining camps saw an opportunity in the fast-growing population of fellow immigrants and began to herd goats and make cheese and dairy products for these growing communities. Among the most well-known of this generation of Utah's cheese entrepreneurs was Luigi Nicoletti, a fifth-generation cheesemaker from Calabria in southern Italy, who arrived in Tooele, Utah, in 1916 and began making cheese from goat's milk. He later moved his herding and dairy operation to Butterfield Canyon, closer to Salt Lake City. In a 1971 interview, Nicoletti recalled a fellow goat rancher who also sold goat's milk cheeses in the Bingham

Canyon area, driving a wagon through the streets and calling out to potential customers "ricotta, formaggio [cheese], crapa [a Southern Italian term for female goat]." Sources mention as many as five Italian cheesemaking operations scattered in the mountains west of Salt Lake City in the early decades of the twentieth century.[27]

Meanwhile immigrant entrepreneurs in California were also herding goats and making cheese. In the Ojai Valley north of Los Angeles, Lorenzo Andreoli kept a herd of five hundred goats and produced cheese which, according to the *Los Angeles Times*, was shipped as far as New York and Chicago. To the north in Shasta County, near the Oregon border, Giovanni Ammirati accumulated a substantial herd of mostly Angora goats (reports range from three hundred to over five hundred) and produced cheese as well as mohair. Ammirati sold goats for meat as well as dairy products to the large Italian community in the San Francisco Bay Area; he advertised his cheese and meat products in San Francisco's Italian newspapers. During the early twentieth century, immigrant cheese production, using both goat's milk and cow's milk, was a significant but little noticed industry in the United States.[28]

As interest in cheesemaking grew within the dairy goat community, established breeders began to recognize the need for scientific research into goat's milk and cheese. Massachusetts native Winthrop Howland was a tuberculosis sufferer who, like many, moved to Southern California for the palliative climate. Howland and wife Martha took up goat raising and turned it into a business, eventually acquiring what they called the largest herd of Toggenburg goats in the country. The Howlands' El Chivar Ranch in Redlands, California, sold goats to Alaska, to missionaries in China, and to the US Agricultural Experiment Station in Puerto Rico. Winthrop Howland was an active member of the dairy goat community: he served on the board of the American Milk Goat Registry Association and judged goats at various fairs and competitions. The Howlands' enthusiasm and tireless marketing also contributed to the elevation of the Toggenburg goat to the status of the most popular and sought after dairy goat of the era.

In 1913, Winthrop Howland had the idea to donate four purebred Toggenburg goats to the University of California's new agricultural satellite campus in Davis, California, known at the time as the University Farm School, or more colloquially, "the Farm." The Farm grew over the years and eventually became a full-fledged institution, the University of California, Davis (or UC Davis), in 1959. At the time the Farm opened, however, little was known about goats and their care, despite the fact that considerable research attention and dollars were already being devoted to cattle, pigs, and sheep at universities across the country. Edwin Voorhies, a new professor at the Farm, was dispatched to Southern California to retrieve the donated goats from Howland's Ranch in Redlands. Reflecting on the experience in an unpublished memoir, Voorhies recalled the faculty's bewilderment: "none of us knew any more about milk goats than we knew about polar bears!" Even so, interest about goat raising and milk was surging both in the state of California and across the nation. The idea to initiate the formal academic and scientific study of goats was a smart, and ultimately necessary, move for the goat industry. The Howlands' timely donation subsequently launched Voorhies and his colleagues on a series of studies of the feeding habits and milk production of goats.

For over ten years, the Farm kept a herd of Toggenburg goats and the faculty devoted itself to goat studies. Voorhies authored one of the nation's early scientific publications on goats, *The Milk Goat in California*, in 1917. But despite his many contributions, Voorhies was at best a reluctant goat scientist. Reflecting on his experiences, Voorhies wrote, "The milk goat venture proved somewhat puzzling. For years jokes came my way, mail arrived in bundles." Despite the skepticism, Voorhies's efforts at the Farm effectively set the school on a path to becoming a national leader in the field. The UC Davis Department of Animal Science remains one of the most prominent educational institutions in the country engaged in the study of dairy goats.[29]

As part of its broader agricultural research operations, the Farm eventually opened a cheese production facility. Due to shortages of all kinds of imported products in the years following World War 1, including cheese, scientists from the United States Department of Agriculture had been studying methods for producing a variety of European cheeses, including Camembert and Roquefort, domestically. Representatives from the USDA traveled to the Davis campus and conducted experiments making blue cheese with goat's milk. With their assistance, the college produced a circular titled *Manufacture of Roquefort Cheese From Goat's Milk*, which outlined the necessary processes and equipment needed to produce the cheese.[30]

As a direct result of this research, a goat's milk blue cheese industry began to blossom on the West Coast during the 1920s. In Falls City, in the foothills of the Oregon Coast Mountain Range, Fannie and Jay Branson began making a Roquefort-style cheese in partnership with Albert Teal, also a goat fancier who had already been experimenting with making cheese from his goat herd. The two families accumulated a substantial herd of Saanen, Toggenburg, and Nubian goats; the herd was said to have

Working in cooperation with USDA scientists, researchers at the University of California Farm (later UC Davis) developed a recipe for Roquefort-style blue cheese made from goat's milk. University Archives Photographs, 1915–1980, Archives and Special Collections, UC Davis Library.

numbered as many as four hundred animals. The cheesemaking took place in an elaborate setting, described in newspaper reports as a series of four interconnected buildings and an aging cave, all built into a hillside and kept cool by the flow of natural springs. The farm's Roquefort cheese was said to have been shipped as far as New York. A local newspaper enthusiastically announced the area's "milk goat boom," and forecasted the advent of an "immensely and enormously profitable new industry." Across the Columbia River, in Carson, Washington, Michael Montchalin of Chevrock Farm also produced Roquefort cheese with milk of his herd of Toggenburg goats. Montchalin, who had grown up with goats in his native France, had big dreams for the future of blue cheese production in the United States; he hoped to establish a large factory in nearby Portland, Oregon, buying milk from goat dairies in the surrounding area and increase his cheese production volume substantially. Unfortunately Montchalin's dreams never came to fruition, however.

By the mid-1920s, the Ettersburg cheese factory in Humboldt County, California, also jumped on the Roquefort bandwagon. The marketing potential of a Roquefort-style cheese became attractive to the group after sales of its Eurisco cheese had stalled. "Most people ... pronounce [Eurisco] a delicious product," said a local newspaper. "But this group of cheese consumers is a small minority and therefore it takes a lot of expensive propaganda to get the new cheese moving in the markets." The company hoped consumers would be more likely to buy Roquefort-style goat's milk cheese because it was a product they were already familiar with. Interest in blue cheese production on the West Coast continued well into the next decade; Mt. Lassen Goat Dairy won a Gold medal for its Roquefort at the California State Fair in 1936.[31]

❖

The rise of goat's milk's reputation as a healthful, protective food jump-started a goat's milk industry in the United States. Enthusiasts developed substantial and productive herds and goat dairy businesses large and small spread across the country. Goats became increasingly popular in the minds of the American public and fostered enthusiasm for goat's milk and goat's milk cheeses, which in turn sparked an expansion of American minds and palates. But goat's milk enthusiasts would soon become acquainted with the economic realities of the Great Depression and World War II, which would dampen their hard-won momentum. Soon the goat dairy industry would redefine itself for the needs of succeeding generations.

CHAPTER FIVE
Back to the Land
A Dairy Goat Renaissance

> *I keep straying on mental anger warpaths,*
> *and then come back to milking the goats.*
> Allen Ginsberg, letter to Gary Snyder, July 1968

In 1945, *The Goat World* magazine published an editorial that confronted a difficult issue: the continuing viability of the goat dairy industry. The essay was unsparing in its assessment: "*The Goat World* realizes that there is something radically wrong somewhere, or otherwise the milk goat industry would have made greater progress in the last few years as its merit warrants . . . just what the reason for its not becoming more popular with the public generally, has caused us to ponder from time to time." This was not the first time industry leaders had pondered their future. As early as 1922, columnists in *The Goat World* addressed the industry's uneven trajectory. There had been no easy answers then; in the 1920s, a post–World War I economic depression and a lack of intra-industry cooperation, among other things, were cited as the most significant obstacles to more widespread commercial success. By the early 1940s, *The Goat World* was itself fading in importance, its dwindling page count a metaphor for the state of goat dairying in the country. The publication folded entirely in 1947. Public enthusiasm for goats and their milk appeared to be waning.[1]

The dwindling interest in goat dairying was not entirely a surprise. As the twentieth century progressed, the human tuberculosis infection rate declined as public officials devoted increasing attention to public sanitation and hygiene practices. At the same time, the efforts of the United States Department of Agriculture's National Tuberculosis Eradication Program to stop the spread of tuberculosis in cattle by testing and culling infected animals had proven largely successful. The threat of dangerous tuberculosis milk, which drove the early growth of the goat dairy industry, had largely been eliminated.

In addition, by mid-century, scientists had isolated a substance called streptomycin, one of a new generation of now-familiar therapies called antibiotics, which could successfully treat certain types of human infection. Streptomycin became the first effective treatment for tuberculosis, effectively ending the long-running epidemic in the United States. Once a cure was introduced, tuberculosis sanitariums across the country closed down as the disease faded from public consciousness. As tuberculosis receded as a significant public health threat, the goat dairy industry's Achilles heel was revealed. Positioning goat's milk as a product primarily tied to an epidemic disease was not a sustainable marketing strategy once the prevalence of that disease diminished.

Meanwhile the advent of World War II diverted public attention and national resources toward the ongoing war effort. As the public became immersed in global events, goat enthusiasts made a number of arguments for their own continued relevance. Many suggested that goat's milk represented a potential solution to widespread shortages and food rationing. Some argued that the war would force Americans to look inward and become more self-reliant, and goats were one potential gateway to self-sufficiency, as back-to-the-landers had already been saying for decades. Others pushed what they called a "victory buck" program, offering male goats to interested farmers as a means to inject high-quality dairy genetics into herds to increase overall milk production. One author even suggested that goat farming could be the perfect antidote to shell shock (now known as PTSD) experienced by returning soldiers.[2]

The goat dairy industry's existential crisis might also be attributed in part to outsized expectations. Gilded Age entrepreneurs such as John D. Rockefeller, Cornelius Vanderbilt, and Henry Ford had all made staggering fortunes in a variety of industries like oil refining, steel manufacturing and automobile production during the late nineteenth and early twentieth century. As a result, hustlers and speculators of the period were always on the hunt for the next big thing, which would make them millions. By the mid-twentieth century, it was abundantly clear that no fortunes were forthcoming from the sale of goats or goat's milk. In part this was because over the decades of its rise, one of the goat dairy industry's biggest difficulties had been a lack of supply. Goat breeders and dairies had simply never developed a sufficient milk supply to meet accelerating demand. The industry's long-standing prioritization of European dairy breeds effectively limited milk production capacity and concentrated goat's milk production in the hands of a few enthusiasts who could afford the top-of-the-line breeds. If, in some altered universe, goat's milk had been able to compete with cow's milk in volume and public visibility during its peak window of opportunity in the late nineteenth and early twentieth century, goat enthusiasts could *perhaps* have won over more hearts and minds to their side of the milk business.

Despite the hand wringing over the future of their industry, all was not lost. The first wave of goat dairy entrepreneurs had not labored in vain—a goat dairy renaissance was imminent.

❖

Kenneth and Cynthia Bice moved from Los Angeles to rural Sonoma County, California, in 1963. They purchased a few acres in Sebastopol, just west of Santa Rosa, a small town with a population of around 2,000 people. The region had once been known for its apple orchards; during the 1940s, thousands of acres across Sonoma County were devoted to apple cultivation and one popular local variety, the Gravenstein, was sent to troops overseas. But by the 1960s, the local apple industry was in decline, and the quiet rural community was the perfect place for a family looking to start a small-scale farm. No one knew then that Sonoma County would become one of the major centers of goat dairying in the United States.

Jennifer Bice remembers that before relocating to Sonoma County, her parents had studied Ed and Carolyn Robinson's *The "Have-More" Plan: A Little Land—a Lot of Living*, a homesteading guide first published in 1943. She still has her parents' copy of the book. *The "Have-More" Plan* is one of a long line of back-to-the-land books and resources that stretch back to the 1864 book *Ten Acres Enough*, a pointed response to rapid urbanization and industrialization in the United States. Over the succeeding generations, others promoted similar ideals. Among the more well well-known was Ralph Borsodi, whose Depression-era critiques of industrialized society in several books included the aptly titled *Flight from the City*, which advocated radical independence and self-sufficiency. Other influential voices were Helen and Scott Nearing, who published *Living the Good Life* in 1954 and went on to establish farms in Vermont and Maine. The principles of hard work, voluntary simplicity, and self-sufficiency espoused by Borsodi, the Robinsons, Nearings, and others appealed to a postwar generation that wished to escape the growing constraints of urban life.

As the Bice family established themselves, goats became a mainstay of the growing family and farm. Mom Cynthia Bice once told a reporter that the family had purchased their first goats because, with five children at the time (the Bice family eventually grew to nine children), "the cow milk bill began to soar alarmingly." While she had at first worried that the children wouldn't like goat milk, it turned out they liked it just fine. The Bice children, particularly Jennifer, the oldest daughter, took on the goats as a 4-H project, and the herd gradually began to expand. Jennifer remembers that "people saw we had goats and would stop by the farm asking if we would sell them goat's milk. It kind of grew from there." The Bice family eventually constructed a milking parlor on their property and first became licensed as Redwood Hill Farm Goat Dairy (later Redwood Hill Dairy and Creamery) in 1968.[3]

Counterculture Goats

In the years following World War II, a series of high-profile events rocked the national consciousness: Rosa Parks's arrest in 1955, conflicts over the integration of segregated schools, and the assassinations of leaders like Medgar Evers and Martin Luther King Jr. The increasingly vocal civil rights movement fought for equality within a society that vehemently resisted it. At the same time, political turmoil in Southeast Asia began to attract the attention of the United States, and by 1967, there were over 500,000 US troops stationed in Vietnam. Protests against the Vietnam War escalated as numerous violent atrocities were broadcast on television. The onset of the draft, which conscripted young men into the military further fanned the flames; many refused to register and some fled to Canada. Campus protests spread around the country as students voiced their discontent the state of the nation. A progressively disaffected and vocal American populace took to the streets, believing there had to be something better.

Some of the disaffected followed Harvard professor and later psychedelic drug advocate Timothy Leary's advice to "turn on, tune in and drop out." Dropping out meant abandoning the strictures of contemporary society and starting over, ushering in the era of the counterculture commune. By 1970, *The New York Times* reported the existence of nearly 2,000 communes in thirty-four states. An estimated 250,000 people, most of them young, white, and middle or upper class, are estimated to have lived for some period in rural communes, along with as many as 500,000 more in urban areas across the country.[4]

Among many other things, communes became ground zero for a wholesale reconceptualization of food consumption in the United States. Once self-styled hippies checked out of mainstream society and hitchhiked their way to a rural outpost, propelled by disaffection and chanting fragments of eastern philosophies, intent on sowing the seeds of radical rebellion, they faced the imperative task of feeding themselves. Robert

Houriet, journalist and author of *Getting Back Together*, a chronicle of his journey across the country visiting communes published in 1971, once said that the main conversational topic on communes he visited was not politics, sex, or the Vietnam War, but food. Whatever conversations or conflicts the intellectual and political philosophies of the day produced, the need for food remained and always returned.[5]

The counterculture approach to sustenance was consistent with its politics. For many, the ultimate goal was to live entirely free from the influences of polluted, pesticide laden, mass-produced food that, as Rachel Carson's *Silent Spring* had shown convincingly, sickened people and the planet. The goal was to return, quite literally, to a more foundational way of living. The earlier generation's systematic *"Have-More" Plan*, with its detailed plans and illustrative charts, gave way to what the Morningstar commune in Sonoma County termed "voluntary primitivism":

> Conditioned by their fast, competitive culture to unnatural living rhythms, Americans find themselves falling sick and dying prematurely in a dis-eased society ... why not explore our common ancestral heritage—a simple shelter, a garden, some goats or a cow, some chickens and plenty of fresh air and sunshine? You don't need more than that to be happy and if you have more it'll probably make you sick.

Stephen Gaskin, leader and founder of The Farm in Tennessee, a spinoff group of Morningstar, preached something similar, an approach neatly summed up in the title of his essay, "This Country Needs in Great Numbers to Become Voluntary Peasants." Hippies moved back toward nature in order to rewrite their future.[6]

While counterculture commune dwellers had big ideas about transforming society, a variety of pamphlets, handbooks, and publications, both formal and informal, soon materialized offering practical advice. The first *Whole Earth Catalog* was published in 1968, and the inaugural issue of *Mother Earth News*, with a print run of just 3,000 copies, appeared in 1970. These and other like-minded publications soared in popularity; the *Whole Earth Catalog* sold 1.5 million copies of its 1971 edition, which won a National Book Award, and within a few years the circulation of *Mother Earth News* skyrocketed to over one million copies per issue. Another prominent periodical, *Rodale's Organic Gardening and Farming* (later shortened to *Organic Gardening*), founded in 1942, found its niche during the counterculture years. By the early 1970s, the magazine had accumulated

700,000 subscribers who wholeheartedly embraced the idea of growing wholesome food with organic principles.

The ever-expanding variety of resources fueled the generation's aspiring farmers and goat keepers. While the first *Whole Earth Catalog* did not include information about raising goats, the second and later issues remedied that omission. Writer Nancy W. Bubel debuted a column called "Notes by a New Goat Keeper," in *Organic Gardening and Farming* in 1971. Bubel's essays chronicle her family's homesteading journey as they first purchased, and then began milking, two goats. "The best thing we learned in those first few days is that goat's milk is delicious," she said. Bubel's columns provided specific details about milking techniques and other tasks, including growing soybeans, which provided fodder for the goats as well as food for the humans. Her easygoing, upbeat stories and advice also detailed the family's other farm pursuits, such as raising chickens and growing their own hay. Bubel eventually became a prolific author, writing several full-length books on farming topics, including *The New Seed Starters Handbook* and *The Country Journal of Vegetable Gardening*.[7]

Not to be outdone, *Mother Earth News* also offered goat keeping advice. The second issue of *Mother Earth News* reprinted Ed and Carolyn Robinson's *"Have-More" Plan* in its entirety, including their advice on keeping goats, marking an implicit handoff of the Robinson's ideals and resources to a newly motivated generation. The handoff, such as it was, did not go as planned: in the third issue *Mother Earth News* editors apologized for the Robinson spread in the previous issue due to reader complaints about some of the Robinson's practices, particularly their recommendations about using chemical sprays for plants. The magazine also received complaints from vegetarians who objected to the Robinson's detailed plans for goat and cow-keeping. "There are many paths to the Clear Light" noted the editors, taking a conciliatory tone; clearly the new generation of counterculture hippies were cut from a different cloth. *Mother Earth News* number six featured Nancy Pierson Ferris's article "Get Your Goat." The author wrote in great detail about the practical ins and outs of acquiring a goat, housing it, and milking it. "I know of no reason why you can't raise a goat or two on your homestead for the same return," said Ferris enthusiastically. For aspiring goat keepers or those simply exploring the parameters of self-sufficiency, an ever-expanding set of honest and reliable resources provided facts, advice, and inspiration and led the way toward new avenues of agricultural pursuit, including goat keeping.[8]

A number of communes of the era kept goats. Author Lucy Horton traveled to forty-three communes across the United States and Canada over a period of eight months in 1971, collecting recipes for her now classic book *Country Commune Cooking*. In the course of her travels, Horton said that "dairy goats are one of the two most popular commune animals, chickens being the other." In many ways, goats were an ideal vehicle for the emerging neoprimitivist commune lifestyle. For those who wanted milk, goats were easy to acquire and smaller and less threatening to farming neophytes than cows. Perhaps more importantly, goats and their milk already possessed an alternative health halo carried over from the early twentieth century, which was perpetuated by a new generation of alternative health gurus including Adelle Davis and others. Davis may have won over the counterculture generation because of her experiences taking experimental LSD in the early 1960s. Horton remembers, "so many communes I visited had a copy of Davis's cookbook *Let's Cook it Right*. The recipes weren't very good," she recalled, "but Davis had already been raising the alarm about the American diet." Though Davis's influence has since faded, she had an enormous impact on this generation's view of food and health, and she recommended goat's milk, especially for infant feeding.

The relationship between hippies and goats was often complicated. Goats exposed a fissure in the midst of the counterculture pastoralist revival between the ideal of self-sufficiency and its practical realization. For some, the very idea of raising animals was repulsive; communities such as The Farm in Tennessee were strictly vegan and did not keep animals or consume animal products at all. In contrast, Robert Houriet, author of *Getting Back Together*, attended a wedding feast at the New Buffalo commune in New Mexico that included barbecued goat, an "electric punch," and an accompanying rock band. Others forged a middle ground, consuming milk but not the animals that produced it. Lucy Horton says she saw a pronounced difference in the approach to animals and meat in communal living arrangements across the nation; she noticed more livestock keeping and meat eating in the East, particularly New England, than she did on the West Coast, where she says animal husbandry was "generally not as big of a thing."[9]

Though goat's milk and meat represented a potential food supply to those striving to live off the land, some hippies struggled to appreciate goats as a food source. When Beat poet Allen Ginsberg bought property on East Hill in Cherry Valley, New York, west of Albany, friend Gordon Ball became the de facto farm caretaker and goat milker while Ginsberg traveled around the world. Ball recalls the skepticism of residents and the many people just passing through. "I drank the goat milk—but I may have been the only one," he said. "Some didn't like the smell." Judith Margolis, who lived at Magic Forest Farm in southern Oregon during the early '70s, was fond of the group's two goats, Zeba and Zora. But not everyone understood the animals or the ins and outs of dairying; some members of the collective did not know that mammals like goats must give birth in order to produce milk. When Zora eventually kidded (gave birth), she didn't end up giving much milk. This generated some internal debate regarding human access to the milk. "Nursing mothers had priority," Margolis said, "If you got up early enough you might get to have some milk in your coffee." But the milk issue was only the beginning. Zora gave birth to a male kid, which led to protracted internal debates about what to do with him. Some members of the group feared that keeping what would eventually be a full-grown male goat was more than they could handle. "The issue of whether or not to slaughter the male goat generated a lot of hostility between two of the male 'leaders' of our leaderless commune," said Margolis. And the vegetarians among them couldn't stomach the prospect at all. The group eventually hired an experienced woman who came to the property with a large knife and slaughtered the goat. She took some of the meat as payment for her services.[10]

Then there was the question of actually caring for goats. Commune farming practices often tended toward the improvisational, as the population of mostly white, middle class young people flocking to communes were oblivious to the nuts and bolts of farming practices. Many let farm animals roam freely indoors and out: Raymond Mungo reported that though Rosemary, a resident goat of Total Loss Farm commune (also known as Packers Corners) was "soft and loving, with helpless eyes," she had to be relegated to living outdoors because of her "biological tendency to shit and piss without regard to carpets or bedspreads." In *Home Free Home*, a collection of reflections written by residents of Morningstar Ranch in Sonoma County, California, Pam Hanna recalled that people brought all sorts of random animals to the property, including goats. Hanna learned to milk one of the goats, named Rosie, from an anonymous Irishman at the farm with a heavy brogue. Rosie ranged freely and was known to occasionally stray into the main house. "She really thought that it should be hers." Morningstar founder Lou Gottleib remembered that the resident goats "taught us a lot," including the basics of goat care such as hoof trimming and dehorning, and were "wonderful to have on acid trips." But Gottleib eventually gave mother goat Rosie and her buck kid Govinda away because few (including, apparently, Gottleib himself) were willing to take responsibility for milking duties. In at least one case, area hippies evaded the livestock question altogether by purchasing goat's milk from a neighboring farmer. Goat farmer Grace Bauder had happily raised and milking goats for decades in Humboldt County, California. As hippies began to populate Northern California in droves in the late 1960s and early '70s, Bauder found that she couldn't keep up with the new wave of demand for goat's milk. She told a reporter that she wasn't at all bothered by the hippies, noting that "their money's as good as anyone else's."[11]

Michael Gies seated on a stump next to Rosemary the goat, Packer Corners commune, October 1969. Photo by Peter Simon (1947–2018), Peter Simon Collection (PH 009), Special Collections and University Archives, University of Massachusetts Amherst Libraries.

Though some struggled to keep and raise goats, others found that they provided both sustenance and a gateway to a new life. Mary Keehn grew up in a navy family and traveled for much of her childhood all over the United States. She eventually became a marine biology major at the University of California, Santa Barbara, landing more or less at ground zero of the counterculture movement in Northern California in the 1960s. She ended up living on a commune outside of Ukiah. "We were staying in teepees and tents there. Somehow, we thought that the land would magically get paid for!" she remembered. Keehn later moved to Sonoma County, where she lived in a barn converted to a dance hall. Having just given birth to her first daughter, Keehn was searching for an additional source of milk, and noticed two goats consuming brush at the farm next door. Keehn asked the neighbors, who ran a cow dairy but used goats for brush control, if she could have the goats. The neighbor responded, "Honey, if you can catch them you can have them!" The two goats, which Keehn named Hazel and Esmerelda, ignited what became a lifelong interest in goat keeping. Keehn eventually became a well-known breeder of champion Alpine dairy goats before starting a cheesemaking business, Cypress Grove Chevre, in Humboldt County, California, in 1983.

Meanwhile in Albion, California, south of Mendocino, a group of women formed a commune and collective called Country Women, a group *Ms.* magazine affectionately termed "Barn Building, Fence Mending, Goat Raising, Well Digging Women." For a number of years the collective produced a magazine called *Country Women*, which served as a guide to rural life geared to a growing segment of counterculture communes that were women-only spaces, asserting independence not only from contemporary politics but also from patriarchal systems that oppressed women. The periodical devoted a considerable amount of space to goat raising and dairying, communicating the commune's considerable knowledge and accumulated resources to an alternative-minded audience. *Country Women* represented a new generation of women farmers in the tradition of early twentieth-century female goat industry pioneers like Irmagarde Richards (see chapter 3). The collective wisdom of the *Country Women* periodicals was eventually compiled in the book *Country Women: A Handbook for the New Farmer*, first published in 1976. Along with sections on land, carpentry, and gardening, the book devotes 146 pages to detailed instructions about keeping, raising, and feeding goats. Within its pages, goat dairying became not only a source of subsistence but also financial independence and self-reliance in a world that doubted women's abilities and confined them to roles as wives and homemakers. "In the country, there is room for a women's renaissance," the authors wrote, "The space and time is there for a total redefinition of ourselves, our relation to the earth, our relation to each other." Ultimately, goats helped the counterculture to envision a path forward.[12]

A New Health Story

Dr. C. E. (Corl Eber) Leach was a longtime goat dairying advocate with decades of experience. During the 1920s, the Leach family had discovered that one of their sons was sensitive to cow's milk, leading the family to purchase several goats. Leach became involved with the goat dairying community of the period and eventually purchased a small publication called *Milk Goat Journal* from Nebraska publisher Rush Deardourf in 1925. By the mid-twentieth century the publication, renamed *Dairy Goat Journal*, had become the primary resource and chronicle of the dairy goat industry. Dr. Leach's son Corl A. Leach eventually took over the publication and became an industry leader in his own right.

During his many years as the publisher of *Dairy Goat Journal*, the elder Leach was a staunch marketing advocate, often encouraging goat owners to market their dairy products more cleverly and aggressively. In a 1960 issue of *Dairy Goat Journal*, Dr. Leach offered prescient marketing advice to a new generation of dairy goat keepers: "We should hold down any theory that goat's milk is a medicine. We must promote goat milk on its own merits, as a health food." Whether or not anyone noticed or appreciated those words, the elder Leach had made an astute observation about goat's milk and food-marketing trends of the period. In the earliest decades of the twentieth century, consumers thought of goat's milk as tonic for a variety of afflictions: tuberculosis, intolerance to cow's milk among infants, skin ailments, and digestive troubles. Goat's milk had been celebrated for the size of its fat globules as well as for how much it resembled mother's milk. Dr. Leach was suggesting that it was time for the goat dairy industry to move forward.[13]

By mid-century, the parameters of what society considered healthy were beginning to evolve. No longer was health simply a matter of ingesting an exacting amount of vitamins and minerals; a new and improved idea of health, tied to a deepened environmental and social awareness, was emerging in the national consciousness. While you could get vitamins from fruits and vegetables, what good were fruits and vegetables if they had been sprayed with harmful pesticides? How could foods produced in factories, pumped full of additives and marketed to Americans for their ease and convenience, actually be healthy? As it happened, goat's milk was well positioned to meet the needs of this new generation of concerned consumers. The solidly rural origins and associations of goat's milk strengthened the perceived connection between goat's milk and health. Small goat farms like the Bice family's Redwood Hill Farm were clearly operating outside the industrial production cycle that had come to dominate and control the nation's food supply. As a result goat's milk began to acquire an updated reputation aligned with the values of the new era. No longer simply an inoculant against tuberculosis, goat's milk became a desirable health food.[14]

Goat milk's evolving identity became solidified thanks to the influence of a new generation of health experts who embraced goat's milk as part of a healthier diet and lifestyle. Interestingly enough, some of these health promoters had some connection to tuberculosis. Paul Bragg was one of the earliest of the Southern California health fanatics who went on to widespread national acclaim. Bragg suffered from tuberculosis as a child and went on to become a lifelong promoter and one of the leaders of the emerging health food movement. Bragg lived well into his eighties, and he attributed his longevity to daily consumption of goat's milk, among other things. Likewise, Robert Bootzin, more familiarly known as Gypsy Boots in 1960s health-conscious California, had a brother who died of tuberculosis. Boots owned the Health Hut, a health food restaurant that became the center of gravity for all things healthy and alternative in counterculture-era Los Angeles. In his 1965 book *Bare Feet and Good Things to Eat*, a compilation of memories, health advice, and recipes, Boots enshrined milk as one of his list of "vital foods" alongside fruits, vegetables, and seeds. For Boots, the best kind of milk (second only to mother's milk) was raw goat's milk, which Boots said, "is easily digested because it is a naturally homogenized milk, and has a rich, creamy flavor." Among his numerous high-profile health-related exploits, Gypsy Boots made regular appearances on *The Steve Allen Show*, a television variety program. In one appearance he persuaded Steve Allen to milk a goat—the moment serving both as a publicity stunt (one of many) for Boots as well as validation for goat's milk adherents. Moments like these drew attention to goat's milk through the modern miracle of a national TV audience. Gypsy Boots's high-profile health and wellness advocacy helped enshrine goat's milk squarely within the contours of the new health lifestyle and food movement.[15]

Among the most vocal of the mid-century goat's milk advocates was Bernard Jensen, a prominent Southern California chiropractor. Jensen's mother died of tuberculosis when he was young, and Jensen himself suffered from a lung malady for which he received treatment throughout his life. Jensen founded a health resort called Hidden Valley Health Ranch near San Diego in the 1950s. At the ranch, sometimes called a "natural cure sanitarium," Jensen provided a wide range of health services, including informational seminars and outdoor activities. Jensen was also an avid goat breeder who kept a herd of Alpine goats at the ranch and recommended their milk as part of a healthy regimen for ranch guests. Jensen's advocacy straddled both the medical and the health-conscious: Jensen believed, among other things, that goats are "sodium-dominant animals," a quality which, he said, kept joints limber and reduced indigestion in those who consumed the milk. Jensen was also an enthusiastic promoter of Whex, a powdered goat's milk product produced by longtime dairy goat breeder and cheesemaker Melvin Eggers at Briar Hills Dairy in Chehalis, Washington. Jensen said Whex provided "a mixed array of minerals and trace elements in easily digested form." Jensen's book of collected wisdom about goats and the health-bestowing properties of their milk, *Goat's Milk Magic*, goes into great detail about the dietary benefits of goat's milk and the myriad ways it improved both his own life and those he treated. For Jensen, goat's milk was not only a healthy product, it was central to maintaining overall human health.[16]

Another significant factor contributing to the growing popularity of goat's milk during the mid-twentieth century was the meteoric expansion of a new category of food shop: the health food store. While dedicated establishments selling so-called health foods had been around for decades, the category began to come into its own by the 1960s and '70s, as did its radical counterculture cousin, the food cooperative. Both venues attracted those seeking healthy and wholesome (and probably organic) products not available at mainstream grocery stores. As a writer for the *Los Angeles Times* summarized in 1972:

> It was books alongside Rachel Carson's *Silent Spring* and others on pollution of the air and water, chemical additives in food and nutritional deficiencies that made people become more concerned about the food in their lives. People began to wonder and ask more questions about where to go to get foods as free as possible from additives, preservatives. As a result more and more began to frequent health food stores.

The growth and popularity of this new breed of food stores was remarkable: a study conducted in 1975 found 4,500 health food stores nationwide, with 20 percent of those located in California. Industry sources estimated the total market for health food products during the mid-1970s to be between one and two *billion* dollars. The Cooperative League of the United States said there were an estimated 920 food cooperatives and "several thousand" buying clubs across the United States in 1979. It was here that fresh goat's milk, yogurt, and goat's milk cheese found a comfortable home among other health food products like whole wheat bread, bean sprouts, and tofu. The National Food Shop in New York City, for example, was well stocked with "[p]umpkin seeds, additive free Canadian raisin bread, fresh goat's milk and herb teas," when it opened in Grand Central Station in 1971. Health food stores and cooperatives ensured that goat's milk was more available than ever.[17]

In Northern California, the Bice family's Redwood Hill Farm sold their goat's milk to small health food stores and co-ops in Sonoma County and the San Francisco Bay Area. Daughter Jennifer was tasked with making the milk deliveries. "My mom didn't drive, so when I turned fifteen, I got an agricultural driver's license. I did all the deliveries myself." While the family focused on milk sales, they also began to make and sell yogurt as well as a then unfamiliar fermented milk product, kefir. By 1972, the Bice family had accumulated a herd of 100 goats. A local newspaper article about the family's successes announced, "The Dairy Goat Business is Booming."[18]

Cooperation is the Solution

Joanna Guthrie Smith and her husband Henry founded a group they called the Phoenix Academy of Cultural Exploration and Design in Chicago in the late 1960s. Henry Smith, a dermatologist by training, was a past president of the Theosophical Society, a late nineteenth-century belief system that informed the couple's philosophies. The Smiths held Phoenix Academy meetings in their home in Wheaton, Illinois, a Chicago suburb, which attracted local young people and college students. Eventually the Smiths sought a means of turning their philosophies into practice and purchased 220 acres of land in southwestern Wisconsin in order to start their own cultural experiment. The group's plans were broader than simply establishing a commune; Joanna Guthrie Smith told a community-planning meeting in Crawford County, Wisconsin, that the group's goals included the transformation of the county into an "idealistic position in the scheme of modern living." The Smiths' "design for living" program envisioned involving the entire community in a variety of improvement programs that would revive the county's economy and improve its residents' way of life.[19]

Michael Hankin, then a Chicago college student, attended a few of the Phoenix Academy meetings and later moved to what became known as Fellowship Farm. "At the time I was looking to get away from everything—the city, the war," he said. New Hampshire native Kathleen Piper, a University of Chicago student who had also become involved with the Phoenix Fellowship, moved to the Wisconsin farm as well. Fellowship Farm functioned loosely as a commune and a three hundred-acre working farm with chickens, goats, sheep, and later cows. Residents worked toward their own self-sufficiency and using some of the farm's products, including goat's milk, for commercial sale. Piper remembers milking the goats and making cheese and yogurt. "We made a Neufchâtel type of cheese. We'd bring the milk in, strain it, add buttermilk, and let it sit." The group also produced yogurt and a smoked cheese and sold many of their products to the Common Market, a food co-op in Madison, Wisconsin, founded in 1970. Both Hankin and Piper left the Phoenix Academy's Farm after a few years; Piper later milked goats for Wisconsin farmstead cheesemaker Anne

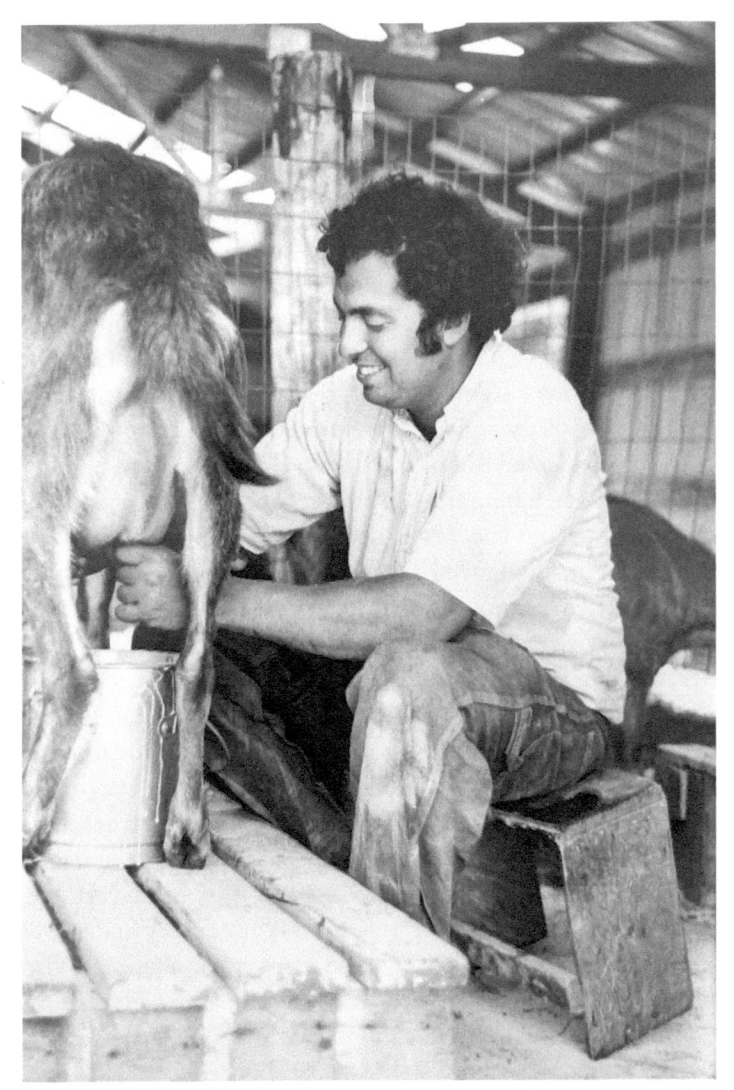

Michael Hankin milking a goat during the 1970s. Photo courtesy of Michael Hankin.

Topham. After the Smiths divorced and sold the farm, Joanna Guthrie moved her cultural improvement project to Madison and started a bakery and restaurant called the Ovens of Brittany in the basement of a health food store called Concordance Natural Foods.[20]

Meanwhile, Michael Hankin managed a herd of goats down the road from Fellowship Farm in partnership with another Chicago expat. As the herd began to grow, they found they had quite a bit of surplus milk, so Hankin approached area cheesemakers about the possibility of using the milk to make cheese. Cheesemaker Floyd Dobbs of the nearby North Clayton Cheese Factory agreed to make cheese for them once a week. Because the factory was set up to handle large volumes of cow's milk, making goat cheese took some improvisation. "He took a smaller vat and rigged it up with steam pipes and experimented with the milk," Hankin recalls. "And he didn't have a pasteurizer, so we made raw milk cheese." The surrounding community took notice of the small cheesemaking operation, and some saw an opportunity. Hankin remembers that "at the time a lot of people were moving out [to rural western Wisconsin] from the Chicago area. "People would stop by and say - hey, we just moved here, if we got goats, would you buy our milk?" Sales grew as the group began to sell their cheeses through an emerging cooperative infrastructure in the region. "There was a guy named Bobby Goldman who had a truck, and he would go around and source food from producers and bring it back to Madison. He bought most of our cheese and sold it to co-ops in Madison."

Hankin eventually moved on from goat raising to start a family, but the area's goat farmers, not wanting to let go of a good thing, formed the Southwestern Wisconsin Dairy Goat Products Cooperative in 1977. Hankin later returned as the cooperative's cheesemaker. The cooperative grew steadily over the next decade and gained national attention for its Kickapoo of Wisconsin brand raw goat's milk cheddar. After the first few years of production, the cooperative was making so much cheese that the area co-ops couldn't handle it all. The cooperative eventually struck a deal with Alta Dena, a raw-milk dairy in Southern California, which purchased the cooperative's cheddar and repackaged it under its own label. Bolstered by the increased demand for goat cheese across the country in the 1980s, the cooperative purchased a defunct cheese plant in Mt. Sterling, Wisconsin, in 1983 and re-branded as the Mt. Sterling Co-op Creamery. The Mt. Sterling Co-op Creamery produced a number of goat's milk cheeses including cheddar, feta, and mozzarella varieties until it closed in 2024.[21]

Although agricultural cooperatives have a centuries-long history in Europe, farmers cooperatives gained momentum in the United States after World War I, when wartime agricultural production surges that supplied overseas troops came to a halt. As prices plummeted, farmers banded together, finding economic power in working together collectively. Now decades later, a new generation focused on prioritizing health and self-sufficiency had become interested in raising goats and selling their milk. While the milk of one farm with ten goats did not amount to much, when pooled with milk from a few other farms, the combined volume became more significant and offered a source of potential income. While developing a goat dairy with all of the requisite equipment to process milk and make cheese was expensive on an individual basis, when multiple farmers worked together the prospect was more attainable. As demand for goat dairy products increased and goat farmers began to innovate, commercial goat dairying and cheese industries began to emerge, functioning on a cooperative basis.

Cooperatives organized around the production and sale of goat's milk or cheese began to appear across the country. In 1944, the Ozark Dairy Goats Products Cooperative began operating in Harrison, Arkansas. The cooperative, with forty-two goat farmer-members, hired a Greek cheesemaker to produce "feta and ricotta as well as other varieties familiar to the foreign trade." In 1955, a group of goat dairies in New York State formed the Goat Milk Producers Cooperative, which hoped to organize and develop a milk plant to process and sell the group's milk. In 1968, a group of dairy goat–farming families got together and pooled their milk, sending it to the Monticello Farmers Mutual Co-op Creamery near Cedar Rapids, Iowa, which produced a cheddar-style goat's milk cheese. The group was optimistic about the market potential for their cheese: Dick Sherman, one of the members, told a reporter that he'd found several out-of-state outlets for the cheese. "Our market is actually bigger than three producers can handle," said Sherman. Another group of goat farmers in Waterloo, Iowa, resurrected the old Steamboat Rock Creamery, which had been closed for several years, and turned it into a goat cheese factory in 1972. The proprietors hoped to sell the goat's milk cheese to a distributor in Pennsylvania.[22]

For over forty years, the Mt. Sterling Co-op Creamery in Crawford County, Wisconsin, produced a wide variety of cheeses made from goat's milk. Author photo.

On the West Coast, Bill and Nancy Ulhorn, graduate students at the University of Oregon, started the Oregon Dairy Goat cooperative in 1971. The cooperative's stated mission was threefold: to provide a market for goat's milk, to include low-income persons in the cooperative, and start new goat herds for families. Fees to join the cooperative were reduced for low-income families, and members were required to donate baby goats to new members to help create start-up herds. The cooperative made cheese at a plant in Salem, Oregon, and the product was distributed by Rogue Gold Dairy in Grants Pass, then owned by California's Vella cheesemaking family. According to one report, by 1978 there were eight commercial goat cheese producers in the United States, along with "a half a dozen or farms [that are] producer distributor operations . . . in Southern Colorado."[23]

One of the largest goat's milk cooperatives in the United States was once based in California. During the early 1920s, the Widemann Company, based in Northern California, had produced evaporated goat's milk in cans (see chapter 4). After the company went out of business, the Meyenberg family, which had already been producing evaporated cow's milk in Monterey County, California, took up the production of evaporated goat's milk. When Meyenberg relocated its operations to Ripon, California, in the 1930s, evaporated goat's milk production followed.

The Meyenberg evaporated goat's milk operation fueled the growth of a community of goat dairies in California's Central Valley, which supplied milk to the plant. In 1948, a group of those goat dairy farmers formed an independent cooperative known as the California Goat Dairymen's Association (CGDA). According to contemporary reports, the cooperative's members were motivated by their desire to diversify their income by finding a secondary outlet for their milk. The CGDA subsequently produced and marketed their own independent evaporated goat's milk product sold under the name Miracle Brand. The CGDA was reported to have been producing 4,000 gallons of evaporated milk a week during the 1950s.[24]

One of the most significant effects of the introduction of the CGDA's Miracle Brand evaporated goat's milk was its marketing strategy. In an effort to compete with Meyenberg's evaporated goat's milk product (until then the only evaporated goat's milk product on the market), the CGDA introduced its product into a new breed of food shop, the general grocery store. Until that time, evaporated goat's milk had been sold exclusively in pharmacies and drug stores or directly to physicians in order to emphasize its purported medicinal properties. General grocery stores, so familiar today, were just beginning to appear in the United States in the early twentieth century when C. H. Widemann first introduced evaporated goat's milk

The Meyenberg Company headquarters in Turlock, California. The company was purchased by France-based Emmi Group in 2017. Author photo.

to the marketplace. By moving its product into the general grocery arena in the 1950s, the CGDA significantly broadened the market for evaporated goat's milk—and by extension goat's milk in general—by increasing the availability and visibility of the product. This action helped facilitate goat's milk evolution beyond its purely medicinal associations.

Despite the efforts of the California Goat Dairymen's Association, the Meyenberg company continued to exert considerable influence over the cooperative's members since the CGDA relied on Meyenberg's plant to produce its Miracle Brand product. This latent dependence eventually became a liability to the cooperative. Over the course of the twentieth century, the Meyenberg family company changed hands multiple times, acquiring a variety of additional businesses under its corporate umbrella and taking on a variety of retooled names, such as Meyenberg Old-Fashioned Products. In 1976, the company was acquired once again, this time by a Sacramento real estate developer who may have been more interested in access to one of the company's assets—Foster's Freeze, a Southern California–based fast food chain. The company's new board of directors reorganized the business entirely, shedding the dairy portfolio and closing the evaporated milk processing plant in Ripon, giving the CGDA only two days' notice. The closure represented a potential death knell for the cooperative and its goat farmers. The CGDA acted quickly, suing to prevent the plant's closure as a temporary stopgap measure. Meanwhile CGDA members pooled and sold their milk to Sonoma Mission Creamery in Sonoma, California, where it was dried and packed into one hundred-pound bags and one-pound cans with the Miracle label and sold to health food stores.[25]

The CGDA eventually managed to avert demise by financing and constructing their own independent milk processing plant in the small city of Turlock, California, which opened in 1976. John Jeter, who became general manager of the CGDA in 1979, worked hard to diversify the cooperative's product line to ensure its long-term growth and survival. Among other things, Jeter helped the CGDA develop an ultra-pasteurized fluid milk product with a longer shelf life. The CGDA also briefly contracted with the Peluso Cheese Company in Los Banos, California, to produce a cheddar-style goat's milk cheese. By 1980, the CGDA plant was receiving milk from sixty-five area goat dairies, some 12,000 gallons per week during peak months.[26]

The cooperative's fortunes changed once again in 1985, when the Jackson-Mitchell Company purchased the Turlock plant and retired the cooperative's Miracle Brand evaporated goat's milk in favor of the Meyenberg name and brand. The California Goat Dairymen's Association closed for good. But the Jackson-Mitchell acquisition was not a total surprise to industry insiders. Jackson-Mitchell had acted as Meyenberg's evaporated goat's milk sales agent since the 1930s under a different company name, Special Milk Products. During the 1950s, the US Department of Justice filed an antitrust lawsuit, accusing the Meyenberg and Special Milk Products companies of colluding to fix the price of evaporated goat's milk. While the case was eventually dropped, Special Milk Products subsequently became Jackson-Mitchell (new company, same owners) and continued to act as broker-distributor for Meyenberg. After the new corporate leadership closed Meyenberg's Ripon, California, dairy plant in 1976, Jackson-Mitchell acted quickly. The company purchased Meyenberg's second evaporation plant located in Yellville, Arkansas, at a bargain price along with the Meyenberg trade name, in effect, preserving its own business in the process. After numerous plot twists and turns throughout the course of the twentieth century, the Meyenberg family's original sales agent merged with the Meyenberg family's evaporated goat's milk business.[27]

A Goat Boom

In 1975, *The Wall Street Journal* published an article about an unusual topic (at least for the *Journal*): goats. "There's a goat boom in the United States," wrote the author, "and that's a sure sign that times are tough." There was indeed a goat boom—more specifically a dairy goat boom—in the United States. According to the *Journal*, the numbers of goats registered with the American Dairy Goat Association jumped from 5,000 annually in 1968 to 20,000 in 1974. There were other indicators as well: California officials reported that the number of goat herds enrolled in the state's milk testing program doubled between 1974 and 1975, from 105 to 230 herds. Across the country, goat breeders reported that the value of goats was soaring; one breeder in Northern California said the value of her purebred Alpine goats, known for their milk production, had tripled during the early 1970s. The rapid rise in demand signaled a full-fledged renaissance of goat dairying in the United States.[28]

The goat boom was actually a more complex and interesting story than *The Wall Street Journal* managed to capture. It's certainly true that high inflation and the OPEC oil embargo dominated the headlines during the 1970s, and citizens were concerned about the economy and their own livelihoods. But the wave of interest in goats represented much more than people trying to save money. Past associations of goats with poverty had long since changed: the population of goats was growing because a new generation of back-to-the-landers and hippies living on communes were acquiring and raising goats as part of a broader movement that questioned food quality and its effects on human health and the planet. Goats were easier to care for than dairy cattle and offered an avenue toward achieving a healthier and more self-sufficient lifestyle. Goat numbers were also booming because the goat's milk business was booming, and the goat's milk business was booming because goat's milk had become associated with the emerging ethos of healthy food. Perhaps most importantly, the goat boom was built on the foundation laid by an earlier generation of dairy goat advocates who had already developed an infrastructure of goat breeding, dairying, and product awareness in the United States. The story of the goat boom was the story of goat's milk finally working its way into the modern mainstream.

❖

As the goat dairy industry was expanding, consumer dairy preferences were also evolving. Influenced by the emerging awareness of unhealthy, mass-produced, and chemical laden foods, a large and growing number of consumers were turning toward raw milk as their milk of choice. While the term "raw milk" predates the invention of pasteurization, the process of heat-treating milk in order to destroy dangerous pathogens, raw milk is generally understood to mean unpasteurized milk. Starting in the 1960s, a new generation of health-conscious consumers argued that pasteurization destroyed vitamins, minerals, and other health-giving properties of milk. The only truly pure and nutritious milk, they asserted, came straight from the animal, without treatment.[29]

The raw milk debates carried particular significance for the goat dairy community. Goat milk's popularity had long been rooted in its health-bestowing capabilities. Even after the early twentieth-century association with tuberculosis faded, goat's milk continued to be considered a tonic for stomach troubles, skin conditions, and colicky infants, among other issues. Goat's milk adherents believed that pasteurization, or other types of treatment, including homogenization, destroyed the very components that made goat's milk uniquely healthy.

But the dairy industry was changing. Over the course of the twentieth century, a variety of new technologies, including refrigeration, milking machines, bulk-milk storage tanks, and integrated transportation systems, had transformed milk production in the United States. The industrialization of dairying made milk production more consistent and efficient and made milk more accessible to consumers across the country. Even as hippies and back-to-the-landers were discovering fresh, raw goat's milk on country farms, regulations and technologies for handling and processing milk were evolving all around them. Pasteurization had long since become the most accepted method of ensuring a safe milk supply. Regulations regarding acceptable milk-handling practices were enshrined by the United States Public Health Service in its Standardized Milk Ordinance (now known as the Grade A Pasteurized Milk Ordinance, or PMO), first issued in 1924. Over the years municipalities, counties, and states incorporated federal rules into their own laws, and federal Grade A milk regulations, including pasteurization, became standard practice for the commercial cow and goat dairying industries across the country.

And so battle lines were drawn. A publication issued in 1966 by the United States Public Health Service, *What You Should Know About Grade A Milk*, details the safety benefits of drinking pasteurized milk. Citing the potential for transmission of such afflictions as tuberculosis, typhoid fever, and strep throat via raw milk, the publication argues, "There is no way of making absolutely sure that raw milk will never contain [these disease] organisms," and for this reason, "health authorities agree that all milk should be pasteurized." Similar arguments were put forward in the popular press. "One of the follies of the new back to nature faddism that is sweeping up so many of the young people is the belief that raw milk is superior to pasteurized milk," wrote nutritionist Jean Mayer in his nationally syndicated column "Food For Thought" in 1972. "And [the risks of raw milk] are in no way mitigated by switching to raw goat milk, as some organic communes are now doing."[30]

As raw goat's milk adherents became increasingly vocal, sales of raw goat's milk accelerated across the country. While federal laws do not permit raw milk sales, federal laws apply only to milk sold in interstate commerce, leaving states to regulate dairy products sold within their borders. As a result, states across the country were forced to confront the raw-milk issue. Farmers and consumers found that the parameters of the many state and local level regulations varied widely. While some states imposed inspection and other safety requirements, others took a more stringent approach. After hearing pleas from goat's milk consumers, the Oklahoma City Council voted to legalize the sale of raw goat's milk, but sales were allowed only if purchasers had a doctor's prescription. A similar prescription requirement for raw goat's milk sales remains the law in the states of Kentucky and Rhode Island.

The expanding scope of oversight and regulation directed specifically toward raw goat's milk angered some goat dairy owners. E. R. Wallace of Mystic Lakes Goat Dairy in Redmond, Washington, just outside of Seattle, expressed disdain at attempts to regulate his raw goat's milk sales. Wallace believed that such laws were a way of forcing small farmers to sell milk to big companies rather than building their own businesses. A goat dairy owner in Wisconsin took another course; after being "scared out of the goat dairy business" by state regulators and finding the prospect of meeting state requirements too expensive, she announced that she would give away raw goat milk for free to any sick child that needed it. On the other side of the equation, Virginia's Cove Mountain Goat Dairy sought assistance from state regulators, saying "bootleggers" selling raw goat's milk under the table were driving their legal goat dairy business into bankruptcy. Nevertheless, raw goat's milk production continued to grow in popularity; one contemporary expert estimated that by the late 1970s nearly half of all licensed goat dairies in the nation produced raw milk.[31]

California became one of the main centers of gravity in the raw-milk debates of the period, in part because of Southern California's Alta Dena Dairy, whose activist owners, brothers Elmer, Edgar, and Harold Steuve, were vocal in their beliefs about raw milk and fought hard with state regulators for the right to sell it. Although Alta Dena was a cow's milk dairy, the company purchased and sold raw goat's milk and goat's milk cheese under its own label as well. Alta Dena's primary goat's milk supplier was

Laurelwood Acres Goat Dairy, a large goat dairy that started in the 1940s in Topanga Canyon north of Los Angeles. In 1965, Laurelwood Acres relocated its entire ranching and milk production operation to Ripon, in California's Central Valley, when urban sprawl filtered into Topanga Canyon and the land became far more valuable than the goat dairy. The dairy's move was particularly notable for its speed and precision; the family team managed to move hundreds of goats over three hundred miles in one night using dozens of trucks.

Because Laurelwood Acres sold raw milk, the farm operated for a number of years under the supervision of the San Joaquin County Medical Commission (a legacy of Henry Coit's turn-of-the-century milk commissions) formed specifically to monitor the dairy and its operations. Among the operating requirements were that the dairy's milkers were required to undergo monthly medical checkups and the goats were required to be examined monthly by a veterinarian and undergo blood tests every other month. Although Laurelwood Acres also produced pasteurized goat's milk and cheese, raw milk made up a significant portion of its sales. According to owner Wes Norfeldt, the farm's raw goat's milk sales increased from 35 percent of its production in the early 1970s to 69 percent by 1979. While it remains legal to sell raw milk, including raw goat's milk, within the state of California, farms that do so must meet a series of state-administered requirements, and raw dairy products are required to carry a label warning purchasers of the potential risks.[32]

For a number of years, Laurelwood Acres was the largest goat dairy in California and one of the largest in the country with a herd of over 1,000 goats. Laurelwood Acres was also well-known among the nation's goat breeders for its purebred Saanen goats, a Swiss breed prized for its milk production; many contemporary goat herds still have animals with bloodlines traceable to prized Laurelwood Acres dairy stock. Laurelwood Acres Goat Dairy eventually closed in the mid-1980s. Though sales of raw milk were controversial during the 1970s (and remain so today), enthusiasm for raw milk, and publicity generated by raw-milk debates, contributed to the expansion of the goat dairy industry during the 1970s.

❖

Jennifer Bice's first experiences with goats were as a kid in 4-H. She said that in the mid-1960s, goats were frowned upon as 4-H projects. By 1967, as the goat dairying world was beginning to grow, the newly organized Redwood Empire Dairy Goat Association had formed and began holding regular meetings in Sonoma County. Bice went on to win a first-place ribbon in goat showmanship at the Sonoma County Fair in 1970 and later that same year was named a Dairy Goat Princess of California. Redwood Hill Farm continued to grow as the business became increasingly successful. After years of showing and judging goats under the tutelage of her parents, Jennifer Bice took over the farm with her husband, Steve Schack, when her parents retired in 1978. By the mid-1980s Redwood Hill had over one hundred goats and was selling its Grade A goat's milk to some sixty stores across the Bay Area. The Bice parents' back-to-the-land sojourn in the early 1960s had evolved into a significant commercial endeavor.[33]

The 1960s in America conjures up vivid images of the counterculture period: Woodstock, Vietnam War protests, the Black Panther Party, hippies, and Ken Kesey's psychedelic bus, Furthur. The era has long since become known for its rebellious philosophy and distinctive design aesthetic. Familiar stereotypes of the era aside, there's an argument to be made that the decade should also conjure up images of goats, because it was during this period that goats and goat's milk reemerged emphatically into the American public consciousness. The growth of both the back-to-the-land movement and the counterculture commune were instrumental in initiating widespread interest in a whole new category of wholesome and natural foods, conceived in response to industrially produced food contaminated with unhealthy pesticides and additives. Goat's milk was a natural fit for growing health food aesthetic, and in the process the aura of healthiness that already surrounded goat's milk was redefined and re-clarified for a new generation. Soon the era's fascination with and consumption of goat's milk would evolve into a thriving goat cheese industry.

:CHEZ:PANISSE:

1517 SHATTUCK AVENUE, BERKELEY, CALIFORNIA 94709 :: 548-5525
CAFE OPEN 11:30 AM TO MIDNIGHT :: MONDAY THROUGH SATURDAY :: NO RESERVATIONS NECESSARY

CAFE
SATURDAY MARCH 28, 1981

SALADS
Mixed green salad with garlic croutons, $2.25
Asparagus with aioli, $4.25
Pollo Forte- a spicy chicken salad with hot pepper, sweet pepper, $5.00
Poisson cru- thinly-sliced fresh mahi-mahi marinated with lime and cilantro and served with avocado, $5.50
Sonoma County goat's milk cheese baked and served with green salad vinaigrette, $4.25

FROM THE SEA
Six Pigeon Point oysters on the halfshell with sauce mignonette, $5.50

SOUP
Roasted eggplant soup with red pepper cream, bread & butter, $3.50

FROM THE WOOD PIZZA OVEN
Calzone with prosciutto, goat cheese, mozzarella, & herbs, $8.00
Pizza with eggplant, onions and parmesan cheese, $6.50
Pizza Messicana- hot & sweet peppers, cilantro, Monterey & Jack cheese, $6.50
Pizza with escarole, capers, olives, fontina & mozzarella cheese, $6.50

DAILY SPECIALS
Fettuccine alla Bolognese- a rich meat, tomato, red wine, mushroom, $6.50
Fettuccine with chicken breasts, lemon, garlic and olives, $7.00
Spezzatino di Vitello- veal stew with turnips, carrots & thyme, $7.50
Shark Riviera- baked with artichokes, capers, potatoes, $6.75

DESSERTS
Almond tarte, $2.25
Ask your waiter about today's special desserts

A LA CARTE
Parmesan cheese, $1.50
Anchovy filets, three, $.75
Bread & butter, $1.25
Garlic, $.50

Minimum table service per person, $4.50, 11:30-3:00 & 5:00-10:30.
We accept cash and personal checks only

Chez Panisse began serving Laura Chenel's Sonoma County Goat's Milk Cheese in the early 1980s. Chez Panisse Records, BANC MSS 2001/148, the Bancroft Library, University of California, Berkeley. Used with permission.

CHAPTER SIX

Say Chevre

From Santa Monica to the Napa Valley, from vegetarian restaurants to bastions of nouvelle cuisine, in California nearly everyone is "doing" goat cheese.

Marian Burros, 1982

In 1977, a group of forty Northern California dairy goat farmers, among them a woman named Laura Chenel, banded together to form a cooperative. Led by reporter turned LaMancha goat breeder Barbara Backus, the group pooled their milk and sent it to the Sonoma Cheese Factory, which produced a Jack style cheese for the group. Chenel assumed the role of salesperson. She once told a reporter, "When I went to cheese shops around Northern California the response I got was, 'This is OK, but you should really make chevre.'" The cooperative disbanded within a couple of years, but Chenel's subsequent quest to perfect a technique for making chevre, a style of fresh goat's milk cheese common in France, would eventually become a pivotal moment for the goat cheese industry in America.[1]

Meanwhile, something was brewing in nearby Berkeley, California, and it was not just the coffee in Alfred Peet's corner coffee shop. The name Alice Waters, chef and co-owner of a small restaurant called Chez Panisse on Shattuck Avenue, was becoming familiar to a national audience. Caroline Bates's glowing review of the restaurant appeared in *Gourmet* in October 1975, followed by another enthusiastic review by James Beard in 1978 in his nationally syndicated newspaper column Beard on Food, in which Beard bestowed effusive praise on Waters's "brilliant gastronomic mind, her flair for cooking, and her almost revolutionary concept of menu planning." The reviews raised the profile of both Waters and the restaurant. In 1979, Waters participated in a series of showpiece dinners paired with Louisiana chef Paul Prudhomme. The American chefs prepared a series of meals alongside French and Italian chefs in various venues around

New York City. The event, which garnered extensive press at the time, was reminiscent of the so-called Judgment of Paris of 1976, in which American wines scored favorably in a blind tasting against French wines. Newspapers proclaimed that "American Chefs are the New Rising Stars," and Waters was one of them. A new ethos was emerging—American chefs were demonstrably as good as French chefs, and, by extension, American food could be exceptional as well. The message resonated with diners across the country. Waters would soon bring goat cheese into the culinary conversation.[2]

Around the same time, Helen Allen, a former English teacher, and her husband Dick, a stockbroker, opened the Wine and Cheese Center in San Francisco in 1975, joining a new wave of gourmet retailers opening across the country. Allen had spent time in France while in college and knew her way around French wines and cheeses. She later apprenticed at the Cheese Board in Berkeley, which opened in 1967. The Wine and Cheese Center, which eventually expanded to three locations, stocked a variety of European cheeses alongside an equally impressive collection of wines from California and around the world. During the early 1980s, the shop even carried coffee beans from an upstart coffee roasting company called Starbucks, whose owners counted Alfred Peet as a mentor. Helen Allen was an enthusiastic advocate of artisan cheesemakers and traveled extensively both locally and abroad in her search for the best cheeses and wines. She dazzled customers by featuring a wheel of Sonoma Jack cheese custom cured for the shop and by regularly bringing in two-hundred-pound wheels of Lutzelfluh Emmental cheese made by Swiss cheesemaker Karl Gerber, who produced just one wheel of the cheese per day.[3]

And so when Laura Chenel felt that she had perfected her French-style goat's milk cheeses, she brought them to Helen Allen. Allen, who knew good goat cheese, sent Chenel to Alice Waters, someone she knew would appreciate a good locally made French-style cheese. Waters placed a standing order for Chenel's cheese, and the restaurant began to receive regular cheese shipments. "Laura would ship boxes of goat cheese from Sonoma County on the Greyhound bus, and we'd have to send someone from the kitchen to the bus station to pick up the cheese every day. And every day this big wet box of cheese would come into the kitchen. It was quite something," remembers Joyce Goldstein, then a chef at Chez Panisse.[4]

Waters and her chefs incorporated Chenel's cheeses into a variety of dishes at Chez Panisse, including a goat cheese and leek tart and baked goat cheese stuffed figs. Perhaps the best-known item on the menu that featured Chenel's cheese was the goat cheese salad, which remains on the Chez Panisse menu. The salad is deceptively simple: a small round of warmed fresh goat cheese coated with breadcrumbs, resting atop lightly dressed salad greens. Journalist and cookbook author Janet Fletcher was one of several cooks working the salad line at Chez Panisse during the early 1980s. "Before Laura Chenel's cheese came along, we were using a French goat log for the salad, probably a Montrachet which has no rind. When we couldn't get that we'd use Bucheron and cut off the rind." The salad was a big hit with diners. "[The salad] gave people a chance to try something new; it was a very sexy presentation, the cheese was soft and a little melty . . . and it took off like crazy," said Goldstein.

As to the origin of the goat cheese salad itself, Waters once told a biographer "there may have been baked goat cheese in France, but I don't think anyone ever paired it with a salad." Food critic Gael Green noted rather acerbically in *New York Magazine* that "a melt of cheese on salad greens, isn't really pure California—I tasted my first sauteed goat cheese on lettuce in Paris years ago." New Yorkers had strong opinions about California goat cheese: Mimi Sheraton, food critic for *The New York Times*, referred to hot goat cheese as a "California cliché." Regional rivalries aside, it's no exaggeration to say that the goat cheese salad marked the beginning of an era.[5]

In the decades since the Sonoma County goat's milk cheese salad appeared on the Chez Panisse menu, a certain outsize significance has been attributed to the meeting of Waters and Chenel. The moment should be put into perspective: Laura Chenel did not invent goat's milk cheese production in the United States—not even close—and was far from the first to sell goat cheese commercially. Alice Waters was not the first chef to serve locally produced goat cheese in an American restaurant. And yet Alice Waters, the chef, and Chez Panisse, the restaurant, were rocketing to national prominence precisely at the moment that Laura Chenel was beginning to craft French-style goat's milk cheeses nearby. Chenel's cheeses were exactly the type of product which Waters, a Francophile champion of the use of fresh, local ingredients, appreciated. As if that wasn't enough, the meeting occurred at a cultural moment when the tastes of the American dining public were rapidly expanding, fueled by the enthusiasm of prominent food writers like James Beard, Mimi Sheraton, Marian Burros, and others. Customers who were dining at Chez Panisse and buying cheese and wine from Helen and Dick Allen were equally eager to try locally produced food products like goat cheese, especially French-style-goat cheese, with its mantle of European authenticity. The meeting of Alice Waters and Laura Chenel, artisan producer of goat cheese in the French tradition, became a serendipitous moment that jumpstarted an era.

The French had already taken notice of the evolution of American tastes; in fact, they had played a role in their development. Beginning in the early 1970s, the French government initiated a marketing campaign in the United States promoting its wines and cheeses to American consumers. Their United States marketing agency, Food and Wines From France, initiated an extensive, multiyear "Say Fromage" promotion, which included a flurry of ads and articles in magazines and newspapers across the country. In one promotional campaign, customers were enticed to purchase French products in the hopes of winning such prizes as French cheeses, dinner at a fancy French restaurant, or the grand prize: a trip to Paris. Cheesemongers, food writers, and industry insiders were wined and dined as the French PR machine worked every angle to increase sales of French products in the United States. These efforts contributed in no small part to the shaping and education of American palates and helped set the stage for the rapid rise in appreciation for French-style goat cheese in America.

A Cheese Revolution

Laura Chenel's nascent company, California Chevre, grew quickly. Chenel quit the two restaurant jobs she'd been holding down to support herself and purchased a small building in Santa Rosa, California, which she turned into a cheesemaking facility. Despite the mythology that has since developed surrounding her rapid rise in the cheese world, Chenel faced the same issues as cheesemakers before and since: she borrowed money from friends, made mistakes, and struggled to get by at first. Quality control was a persistent issue early on: "[b]ecause I was so new at making chevre, I didn't always get the process right," she once said. "My errors created interruptions in the process and caused problems with customers." Production levels were very small, especially in the first few years, and only a very few restaurants and specialty shops were able to get their hands on *the* goat cheese from California.[6]

Nevertheless, demand soared. Chenel's sales were reportedly nearly $500,000 annually by 1983. Publicity for Chenel, and by extension goat's milk cheese generally, exploded. Newspapers across the country ran enthusiastic recipes including goat cheese. *The Monthly Magazine of Food and Wine* (now *Food & Wine*) declared chevre "Cheese of the Year" in 1981. Chenel's California Chevre was featured at a state dinner for Queen Elizabeth II hosted by President Ronald Reagan on the occasion of her visit to San Francisco in 1983. Laura Chenel even appeared in *People*, the first (and perhaps only) time a cheesemaker has appeared within the pages of a celebrity tabloid. Goat cheese had *arrived*.[7]

Though Laura Chenel became the default name and face of goat's milk cheese in America during the 1980s, the emerging goat cheese revolution was really much broader in scope than just a single cheesemaker. Across the nation, a new wave of goat cheese producers was already busily fueling the market for goat cheese in the United States. In fact, the main center of goat cheese production during the 1980s was the Northeastern United States, due to the region's large, affluent population centers like New York and Boston. If precise production numbers from this period could be calculated, goat's milk cheese production east of the Mississippi likely outpaced that of California and the West Coast during the 1980s, despite the higher profile of Chenel's California Chevre.

Among the new wave of Northeastern goat cheese producers was Barbara Reed's Little Rainbow Chevre. Reed kept a mixed herd of thirty Toggenburg and Nubian goats on four acres in Hillsdale, in New York's Hudson Valley. Her cheese repertoire included a fresh chevre, feta, several styles of soft ripened cheeses, and a goat's milk blue cheese she called Berkshire Blue. Likewise, Sally and Theodore Wieninger kept seventy goats on their small farm nestled in the Catskills in Hunter, New York. While many goat's milk cheese producers of the period concentrated on fresher styles of cheese, the Wieningers produced aged cheeses. In his book *Cheese Primer*, Steve Jenkins, then a cheesemonger at Fairway Market in New York City, gushed about Sally Wieninger's washed curd gouda-style cheese, comparing it to a Tuscan sheep's milk cheese. Jenkins even went so far as to label the cheese an American Treasure, a term he reserved for the best of the best domestically produced cheeses.[8]

One of the higher profile of the Northeastern goat entrepreneurs was Miles Cahn, who first made his fortune with the Coach leather goods company. In 1983, Cahn purchased a three-hundred-acre property in upstate New York and started a goat dairy and farmstead cheesemaking business called Coach Farm; he sold the leather goods company several years later. In a 2012 interview, Cahn recalled that "we knew that Chenel was selling [goat cheese] out on the West Coast, and we figured we had New York City, the biggest market possible. So it seemed logical and so simple that we embark on [this project to make goat cheese]." In pursuit of creating the best quality product possible, the deep-pocketed Cahn hired a variety of experts, among them French cheesemaker Marie-Claude Chaleix, who had made cheese on her family's farm in France (and, incidentally, was one of the people Laura Chenel studied with in France). Cahn also lured Wes Norfeldt, formerly of Laurelwood Acres Goat Dairy in California, out of retirement to manage Coach Farm's substantial herd of goats. While Cahn sold Coach Farm in 2006, the company continues to produce a variety of goat cheeses.[9]

Another of the emerging Northeastern goat cheesemakers was Vermont Butter and Cheese Company (now Vermont Creamery), started by Allison Hooper and Bob Reese in 1984. The company got its start when Reese, then marketing director at the Vermont Agency of Agriculture, got the idea to serve goat cheese at a state sponsored event. Allison Hooper knew how to make goat cheese because she'd apprenticed at several goat dairies in France while in college. The product proved to be so popular that a goat cheese company was born. Meanwhile, cheesemakers in Maine were also beginning to sell their products;. among that state's earliest wave of goat cheesemakers were Camilla Stege, a self-described back-to-the-lander who made cheese under the Moosetrack brand, Barbara Brooks of Seal Cove Farm, and Penny Duncan of York Hill Creamery. Both Brooks and Duncan took cheesemaking classes from Stege, who says she picked up her cheesemaking skills from an old cheesemaking book she found after purchasing her first dairy goat in 1972.

While goat cheese production accelerated quickly on the East Coast during the early 1980s, goat cheese had already accumulated substantial momentum in the Midwest with the founding of the Southwestern Wisconsin Dairy Goat Products Cooperative in Wisconsin in 1977 (see chapter 5). Several other cheesemakers also started in the region during the 1970s. Jere Linda Sayer and husband Phil produced a cheddar-style goat cheese from the milk of their 200 Goats in New Providence, Iowa. The cheese was sold under the Capricorn brand and distributed nationally and was featured at Zabar's in New York City. In northern Minnesota, the Winger Cheese Company began processing goat's milk along with cow's milk and sold the goat cheese to local grocery stores and via mail order as early as 1975.[10]

Many of the new generation of goat cheesemakers came out of the back-to-the-land movement. Judy and Larry Schad moved their family to southern Indiana in 1976. "I was a Rodale groupie!" Judy Schad said, "I wanted to raise my own vegetables and the whole thing. For a long time it was a comedy of errors, we had no idea what we were doing," Eventually a neighbor recommended that the family get a goat for milking. One goat became two, and Schad began experimenting with making cheese in her kitchen. Capriole Farms grew to become one of the most well-known and loved producers of goat cheese in the nation.

Vincent and Christine Maefsky both grew up in Brooklyn, New York, and got to know each other through their involvement with the Catholic

Worker House in New York City. After graduating from college and working for a year in Oklahoma, the couple purchased a farm in eastern Minnesota and started out with six goats. "We were motivated by the Catholic Worker tenet of cultivation of the land. Growing things, supporting ourselves through our own efforts was something important to us," said Christine. Soon they had people coming to the farm asking to buy raw goat's milk. By 1975 they started to package and sell their Poplar Hill Dairy goat milk to stores in nearby Minneapolis. They sent their surplus milk to Bass Lake Cheese Company in Wisconsin, which produced cheese for them under the farm's name.

The story of Brier Run Creamery in West Virginia was also a tale of New Yorkers seeking a new life. New York teachers Greg and Verena Sava realized their own back-to-the-land dream when they saw an ad in *Mother Earth News* for cheap land in West Virginia—cheaper, at least, than farmland in Vermont, which had been their first choice. The couple moved to Birch River, West Virginia, in 1975 and set out trying to make their living from the land. After trying vegetable farming and raising poultry, among other endeavors, they turned to goat dairying. Brier Run Creamery was one of the first goat cheese operations to become licensed organic in 1990. The Savas made cheese for several decades before closing their business in 2002.[11]

The high-profile debut of Laura Chenel's goat's milk cheeses in the early 1980s drew the attention of consumers nationwide and effectively opened the market for goat dairy products. Across the United States, goat farmers and aspiring cheesemakers saw that goat cheese production could be a viable business model. The stage was set for the rapid expansion of the goat cheese industry in the United States.

The Commercialization of Goat Cheese

For those who aspired to take their cheesemaking to a commercial level during this period, one of the biggest barriers to entry was the relative absence of supplies and equipment for use in making cheese. Smaller cheesemaking vats, properly sized molds, and critical ingredients like rennet and cultures were not easy to find, and of course this was decades before the advent of internet shopping. Barbara Brooks of Seal Cove Farm, who has been keeping goats and making cheese in rural Maine for over thirty years, captured the predicament perfectly when she said, "In those days I'd walk through the hardware store and think, what can I find here that will help me make cheese?" Cheese recipes were also in short supply, and while popular farm and homesteading publications such as *Mother Earth*

News dispensed cheesemaking recipes and advice, that information was not helpful for those who aspired to sell larger quantities, or European styles, commercially.

As a result, some aspiring cheesemakers looked toward Europe. Ian Zeiler, an anthropologist by training, traveled extensively in France with wife Denise Fourant before the pair embarked on their own cheesemaking adventure in Interlaken, New York, during the early 1980s. Allison Hooper of Vermont Creamery spent time as a college student at farms in Brittany and the French Alps, helping manage goats and make cheese. Hooper later apprenticed with Gail LeCompte at Goat Works in Lebanon, New Jersey, another of the emerging goat cheese producers in the Northeast, who made a soft, fresh style cheese called Chevreese. After many of her early cheese experiments failed, Brier Run Creamery's Verena Sava, a native of Switzerland, read about an Alpine cheese inspector in a Swiss magazine and wrote to him for advice on making cheese.[12]

Help was on the way. One of the key players in the growth and development of the burgeoning goat cheese marketplace—and the artisan cheese industry generally—was the New England Cheesemaking Supply Company, started by Ricki Carroll and then-husband Bob Carroll in 1978. The Carrolls, goat owners themselves, had experimented with making various types of cheese in their home kitchen, but found it difficult to do well because it was hard to find the proper supplies. "At that time," said Ricki Carroll, "a lot of goat dairy folks were just dumping their milk because they didn't have equipment to make cheese, and the supplies that were available were difficult to buy in small, goat dairy appropriate quantities."

The Carrolls' foray into the cheese supply business was a bit of an accident. Thinking they might sell a few cheesemaking products such as cheese molds or rennet by mail order, they placed a classified ad in *Dairy Goat Journal* right before they took a trip to England. The ad offered a catalog if responders sent twenty-five cents to cover postage. "When we got back our mailbox was stuffed with quarters," Carroll said. " So we thought, what do we do now?" What they did was start the now well-known New England Cheesemaking Supply Company.

New England Cheesemaking Supply's earliest customer base consisted largely of new and aspiring artisan cheesemakers. Demand for cheesemaking supplies was so strong that the Carrolls quickly sold out of their early inventory, especially rennet, which was hard to find at the time. "We literally had to search around the world to look for suppliers, the demand was so high," said Ricki. The company developed a network of interested customers by attending regional and national goat shows and conferences.

The earliest issues of the company's monthly newsletter, *Cheesemakers' Journal*, reads like a who's who of the emerging artisan cheesemaking industry.[13]

Once prospective goat's milk cheesemakers crafted their products, the next task was to sell them. Penny and John Duncan of York Hill Creamery in Maine benefited from nearby farmer' markets, including the Brunswick Farmers Market, which started in 1977; Anne Topham of Fantome Farm sold her cheeses at the Madison, Wisconsin, farmers market, which opened in 1972. Gail LeCompte of Goatworks sold her goat's milk cheeses at the Union Square Greenmarket in New York City, which opened in 1976. In many parts of the country, however, farmers markets, which provided direct access to an interested public, were not yet up and running. As a result, many of the new wave of goat's milk cheesemakers marketed their products to health food stores and food co-ops, many of which were already selling goat's milk. "The relationships we had already built with health food stores that sold our milk really helped when we later started making cheese," said Jennifer Bice of Redwood Hill Farm. As goat's milk cheese gained popularity, regional grocery chains became interested, and soon goat's milk cheese was available at grocery store chains such as Hannaford in the Northeast and Jewel in the Chicago area. Another significant sales venue for goat's milk cheesemakers was the new breed of higher end gourmet food shops opening in big cities that were eager to feature the latest in locally produced products, including goat's milk cheeses. Among these was Dean and DeLuca, which opened in New York's Soho neighborhood in 1977. Steve Jenkins, who would later become an internationally renowned cheese expert and domestic cheese advocate, developed the cheese program at Dean and DeLuca. According to one report, during its early years Dean and DeLuca carried twenty types of goat's milk cheeses. Barbara Kafka's Star Spangled Foods opened in New York City in 1982; the shop distinguished itself by featuring only foods made or grown in the United States. Kafka, then a well-known food writer, said of her shop, "America's food time has come . . . I think it's time we stood up and were proud of ourselves." Clark Wolf, manager of Star Spangled Foods, had previously worked at Sonoma County's Oakville Grocery, one of the first retail outlets in the country to carry Laura Chenel's California Chevre. Wolf ensured that Star Spangled Foods carried Chenel's cheeses as well. Through an increasing variety of retail outlets, goat's milk cheese found its way into the mainstream marketplace.[14]

Restaurants played a key role in nurturing the growing popularity of goat cheese. Among the high-profile chefs who promoted the product was

Larry Forgione of An American Place in New York City, who featured Laura Chenel's cheese regularly on his menus and credited Chenel with inspiring him to cook with goat cheese. Jonathan Waxman at Michael's in Santa Monica and later Jams in New York City also regularly used Chenel's cheeses, shipped weekly to New York City by express mail. In Los Angeles, Wolfgang Puck put Chenel's fresh goat cheese on pizza at Spago.

Across the country, a new generation of chefs partnered with local cheesemakers. Judy Wicks' White Dog Café in Philadelphia featured Greystone Chevratel, a fresh goat's milk cheese made by Douglass Newbold from the milk of her herd of Nubian goats in nearby Malvern, Pennsylvania. Newbold, herself a chef who studied under Jacques Pepin, had been the executive chef at the 1970s-era Philadelphia restaurant Fish Market before leaving the industry and becoming a goatherd. She said she sold her small output of cheese exclusively to Philadelphia-area restaurants. An entire generation of chefs and restaurateurs were turning away from the structured French formality that had defined American restaurants for decades and moving eagerly toward locally raised meat, fresh produce, and local goat cheese. "My first customer was a friend who owned the Larrupin Café," says Mary Keehn, who started Cypress Grove Chevre in Humboldt County, California in 1983. "She said - I love your cheese and if you get licensed I will buy it." Keehn did become licensed, and her Cypress Grove Chevre went on to become one of the biggest players in the modern-day goat cheese industry.[15]

Bruce Naftaly, chef at Seattle restaurants Les Copains and later Le Gourmand, was an early champion of sourcing local food products in the Pacific Northwest. Naftaly nurtured a number of the early wave of cheesemakers in the region, among them David Greatorex of Kapowsin Dairy. Greatorex, an Englishman, switched from a career as an engineer to goat cheesemaker and started a farmstead operation south of Tacoma, Washington, during the 1980s. "David was the only local cheesemaker who made a soft ripened style of goat cheese. His stuff was fabulous, it would bring tears to your eyes, it was so good," remembers Naftaly. One of Greatorex's passion projects was to produce a goat's milk cheese modeled after the English Wensleydale, first produced by Cistercian monks in Northern England in the twelfth century. Toward that end, Greatorex located a recipe for the cheese in the archives of the British Museum; after tinkering with the details, he produced a Wensleydale-style goat's milk cheese, which he called Capricese, as well as a blue version called Capriblue.[16]

The importance of the relationships forged between cheesemakers and chefs of this era cannot be overstated. Judy Schad of Capriole Farms in

Indiana said one of her earliest customers was Kathy Cary of Lily's Bistro in Louisville, Kentucky. Cary was a local pioneer of farm to table cuisine and has been called the "Alice Waters of Kentucky." Her early menus were inscribed with the phrase "God bless our local farmers." Likewise, Sam Hayward, now of Fore Street in Portland, Maine, has been called "the East Coast's answer to Alice Waters," and no wonder: Hayward nurtured Maine's goat cheesemakers during the early 1980s by buying their cheeses, among them was Barbara Brooks and Seal Cove Farm. Relationships forged between chefs and cheesemakers helped introduce goat cheese to consumers and spurred the growth of the early goat cheese industry.[17]

American Goat Cheese Goes Global

In 2006, French dairy company Rians Group announced that it had purchased Laura Chenel's cheese company. *The New York Times* pronounced, somewhat ominously, "For American Chevre, an Era Ends." What initially appeared to be an isolated business acquisition soon became an economic trend. In 2010, prominent Swiss dairy brand Emmi purchased California's Cypress Grove Chevre (now Cypress Grove Cheese Company). Within a few years, Emmi acquired both Redwood Hill Dairy and Creamery (2015) and the Jackson-Mitchell Company, owner of the Meyenberg brand of goat's milk products (2016), both also based in California.

Then the wave of acquisitions spread eastward. In March 2017, Minnesota-based Land o'Lakes announced that it had purchased Vermont Creamery for an undisclosed amount. Later that same year, Canadian dairy products giant Saputo purchased Wisconsin-based Montchevre for $265 million. Just two years prior, Saputo had also purchased Woolwich Dairy, a goat dairy products producer based in Ontario, Canada, which owned a satellite plant in southwestern Wisconsin. The later Montchevre purchase made Saputo the biggest player in Wisconsin goat cheese. In just over a decade, four large corporations, three headquartered outside of the United States, gained control over the production of the majority of goat dairy products in the country. It was indeed the end of an era.[18]

Sales and acquisitions in the cheese and dairy world are hardly a new phenomenon. Numerous small goat cheese operations have changed hands over the decades, among them Westfield Farm in Connecticut, started by Letty and Bob Kilmoyer in 1971 and sold to Bob and Debby Stetson in 1996, and Belle Chevre in Alabama, which has at the time of this writing changed hands twice since Liz Parnell started Fromagerie Belle Chevre in 1989. Foreign investment in the United States cheese industry was also not new; by the late 1980s there were three French dairy companies operating cow's milk cheese plants in the United States and one, Bresse Bleu in Wisconsin, also produced goat's milk cheeses. Nevertheless, the scope and scale of these more recent transactions had the effect of bursting the idyllic farmstead bubble once synonymous with goat's milk cheese production. Ultimately, as economists might put it, the goat cheese industry was maturing.

❖

What does the new era of "big goat dairy" in the United States look like? One of the most significant and visible changes in the goat dairy industry since 2006 has been an enormous investment in equipment and infrastructure across all of the acquired United States properties. In 2011, Rians announced the opening of a brand new, state-of-the-art 30,000-square-foot creamery in Sonoma County to produce the Laura Chenel brand of goat dairy products. The cost of the project was not publicly disclosed (the new facility has since been remodeled and is now, among other things, LEED Gold certified). Likewise, Land o'Lakes embarked on a reported $10 million expansion of the existing Vermont Creamery facility, which increased the footprint by nearly 40 percent, as well as adding new processing equipment and upgrading packaging and shipping infrastructure. After purchasing Cypress Grove Chevre, Emmi sank $14 million into expanding the existing production facilities to increase and streamline production capacity and an additional $4 million into developing and constructing a new goat dairy. Well over $100 million has been spent in the construction, expansion, and modernization of goat dairy facilities and milk and cheese processing plants across the United States since 2006. The American goat cheese industry of the twenty-first century is modern and state of the art—less farmstead, more stainless steel.

As goat cheese production has exploded in the United States over the past several decades, the dairy goat population, while substantial, has not increased as quickly. Though the dairy goat population has grown steadily since the year 2000, as of January 2024 the total population of dairy goats in the United States was just 415,000. The states of Wisconsin and California perennially top the list of states with the largest population of dairy cows in the nation, and these are the top two states in dairy goat population as well. More recently, Wisconsin has taken a substantial lead over California. In January 2024, Wisconsin was home to 74,000 dairy goats while California's population numbered 37,000.

The Vermont Creamery factory in Websterville, Vermont. The operation has grown considerably since it was founded in the 1980s. Author photo.

These large goat populations begin to make sense when you realize that both California and Wisconsin are also home to the nation's large industrial goat's milk cheese production facilities. In Wisconsin, the goat population is concentrated at the two large goat dairy farms located within a few miles of each other in Chilton, in the northeastern part of the state, an area sometimes called the "Dairy Goat Capital of the United States." The first, Drumlin Dairy, was started in 2017 by Wisconsin-based cow dairy company Holsum Dairies. Just down the road, Chilton Dairy was started by another cow's milk production company, Milk Source, in 2016 (Milk Source operates a string of large-scale cow dairies across the Midwest). While the population of both goat dairies varies seasonally, each houses in the range of over 10,000 dairy goats at any given time. While larger goat dairies with populations of 500 to 1,000 goats or more are becoming increasingly common across the country, these Wisconsin mega-dairies are by far the largest goat dairies in the nation. In early 2022, Drumlin Dairy and Chilton Dairy merged and both are currently owned entirely by Milk Source. The dairies provide most of the goat's milk for nearby LaClare Family Creamery.[19]

Despite the presence of increasingly larger goat dairies, the domestic goat cheese industry is currently operating at a large and growing milk deficit. The population of dairy goats is far less than is necessary to produce the amount of milk being used by the industry; the largest goat dairy products manufacturers in the country have been operating at a milk

Chilton Dairy in Calumet County, Wisconsin, is one of the largest goat dairies in the United States. Author photo.

deficit for years. It's no secret that the producers like Vermont Creamery, Saputo (producer of Montchevre and Woolwich brands of goat dairy products, among others), Cypress Grove Cheese Company, Laura Chenel Company, and others import significant amounts of what is known in the industry as "frozen curd," a partially fermented milk product frozen for later use. There is simply not enough goat's milk being produced in the United States to make the amount of cheese being consumed at current levels. The expansion of goat cheese production in the United States in the past several decades has been underwritten by imported raw materials.

The practice of freezing goat's milk is not an entirely new innovation. Once refrigeration (and freezers) became widespread in the 1930s and '40s, some small-scale goat dairies began freezing a portion of their milk supply, typically more plentiful during the summer, for use during winter months. During the 1960s, goat dairy farmers discussed the ins and outs of the practice in *Dairy Goat Journal*, the leading industry periodical of the day. Jennifer Bice of Redwood Hill Farm once explained to a reporter that frozen curd was commonly used in the industry to smooth out highs and lows in milk production, particularly during the winter months. "[T]o get through the winter we do mix fresh and frozen [curd] together. The flavor is not tremendously affected. The thing that does happen is it's a drier more crumbly cheese. Fresh chevre is very spreadable, very creamy. A frozen chevre product is going to be drier and crumbly," she added.[20]

So where is all of that frozen curd coming from? France-based Emmi, which owns Cypress Grove Cheese Company, Redwood Hill Dairy and Creamery, and Meyenberg, also owns Bettinehoeve, a goat dairy processing company in the Netherlands which also happens to be the largest exporter of frozen curd in the world. Frozen curd also comes to the United States from several other countries, including France, Mexico, and Canada.

The presence and use of frozen curd affects the goat cheese industry in a number of ways. Some goat farmers have accused the larger companies of acquiring cheaper frozen curd from abroad as a means of manipulating the price they pay local farmers for their goat's milk; in effect the frozen product becomes a tool to leverage the price companies pay to farmers for their milk. In addition, because frozen curd is imported from outside the United States, the prices are tied to global currency markets. A strong US dollar in relation to other international currencies means that companies are able to buy more frozen curd from abroad, potentially allowing them to minimize their local milk bill.[21]

Some smaller artisan goat cheese producers are becoming more vocal about the practice of selling goat cheese produced with imported, frozen curd. A grassroots "Say No to Frozen Curd" effort started by Veronica Pedraza of Blakesville Creamery in Port Washington, Wisconsin, pointedly highlights the many inherent contradictions of the issue. "I find it disingenuous that we have companies here in the United States that purport to make American cheese with ingredients imported from a foreign country," says Pedraza. Unlike the larger companies, Blakesville Creamery produces goat's milk cheeses in Port Washington, north of Milwaukee, using milk produced by its own herd of around 2,000 dairy goats.

The potential complications of the big goat dairy manufacturing model came into sharp focus when Canada-based Saputo closed its Montchevre plant in 2022. The plant was a large goat cheese making facility touted at the time as the largest goat cheese production facility in the world. The closure led to the loss of more than two hundred jobs in the small community of Belmont, Wisconsin, which had a population of 1,005 in 2021. Pedraza says, "it's hard to watch those big companies disinvest in agriculture in rural communities in Wisconsin." Mateo Kehler of Jasper Hill Farm in Greensboro, Vermont, echoes these sentiments. "My concern is that the big [goat dairy] companies are not interested in creating a sustainable foundation at a community level. They have essentially substituted industrial commodity curd for real milk that's tied to land, people. All under the guise of artisan cheese."[22]

❖

Going forward, one of the most significant challenges to the goat cheese industry, and the cheese and dairy industry as a whole, is emerging from the fast-growing plant-based "dairy" sector. While goat's milk has been marketed for over a century as exceptionally nourishing and healthy, space in the dairy category is increasingly being claimed by products crafted from almonds, soybeans, cashews, and more recently, genetically engineered dairy proteins. These products are increasing the stakes of healthiness by claiming the additional virtue of being animal free. Traditional cheese and dairy producers are taking notice; many call the plant-based sector's tactics a "war on dairy." The plant-based sector clearly has the traditional cheese and dairy products industries on the defensive.

Early nondairy "cheeses" created for the benefit of the counterculture generation were mostly unappealing rubbery cheese facsimiles. Contemporary nondairy cheese-like products smell and taste pleasantly fermented, nutty, and mushroomy, and some are even appetizing in their own right. Instead of milk, these products are typically made with soy or a variety of plant-based oils. The complex, proprietary chemical formulas used to craft cheese-adjacent products have ushered in an era of what author and cheesemonger Gordon Edgar has termed "remystification of food." Some entrepreneurs are even producing plant-based products marketed specifically as goat-adjacent, enticing consumers with catchy marketing terms like "goatless cheese" and "fauxmage chevre." Whether or not these products represent a significant economic threat to goat cheese producers remains to be seen. But whatever you want to call them, plant-based products crafted to resemble fermented dairy products such as cheese or yogurt continue to improve in quality, and the results are clear: consumers are turning to alternative dairy products, including alternative goat cheese products, in increasingly larger numbers. According to the Food Institute, total sales of plant-based cheeses in the United States amounted to $224.7 million dollars in 2022, and most industry predictions forecast that the category will grow by at least 10 percent or more annually over the next decade.[23]

Plant-based producers are becoming increasingly sophisticated in their quest to craft nondairy products. A new generation of manufacturers are currently producing what are known as bioidenticals, substances that mimic proteins such as casein found in milk. The proteins, manufactured using proprietary processes, are used to craft products that, chemically speaking, are more or less exactly like cheese and other traditional dairy products. Among the companies involved in these efforts is Tomorrow

Plant-based products made to resemble fresh goat's milk cheese are growing in popularity. Author photo.

Farms (originally Tomorrow Foods), which launched its Bored Cow line of "animal flavored milks" made with bioengineered casein produced by Bay Area–based laboratory Perfect Day in 2022. Tomorrow Farms received $10.5 million in venture capital and investor backing, and Perfect Day is poised to become a powerhouse in the casein fermentation space, having secured a whopping $750 million in venture-capital backing in 2021. The nondairy products industry is expanding rapidly across the globe; Fermify, based in Vienna, Austria, seeks to "domesticate microorganisms" in its quest to produce casein via fermentation and "remove cows from the dairy industry by 2027." In 2023, France-based Nūmi announced plans to produce a laboratory-based milk formula for babies. The alt-casein sector is becoming increasingly crowded with well-funded players, all with lofty goals and catchy marketing slogans. The message for the goat dairy industry is crystal clear: the parameters of what the public considers a "healthy" dairy product are rapidly evolving, threatening to leave goat dairy products behind.

❖

The twenty-first century has introduced a new era of big money and international players in the American goat dairy and cheese sector. It's worth asking: is the era of small farmstead goat cheese production effectively over? Increasingly, the answer is emphatically yes. There are a growing number of factors working against smaller producers of goat's milk cheeses and dairy products in the current marketplace. Most importantly, the rise of ever-more stringent food safety laws has meant that dairies, both cow's milk and goats' milk, must implement an ever-larger number of expensive safety measures. In addition, many wholesale distributors have started to require extensive safety documentation, including a full-scale hazard analysis and critical control point plan and third-party audits as a prerequisite for distributing a farm's products, whatever its size. While safety regulations are certainly important, increasing regulation can be prohibitive in scope and cost for a small artisan producer. Small-scale cheesemakers of all kinds argue that many of the new generation of safety laws have been designed around large-scale industrial operations and are not appropriate for a small dairy or cheesemaking facility. Either way, the increasing costs of regulatory requirements effectively forces out smaller players.

In response to these and other issues, some smaller scale farmstead goat's milk cheesemakers have diversified. Some goat farmers have ventured into agritourism, offering regular tours or farm stays; Lively Run Goat Dairy in upstate New York, for example, offers goat yoga during the summer months. Others have broadened their product base; Fraga Farm in Gales Creek, Oregon, sells soaps, lotions, and skin balms, all produced using their own organic goat's milk. Some have begun producing mixed-milk cheeses, combining available goat's milk with other types of milks to make blended-milk cheeses. Goat Rodeo Farm in Allison Park, Pennsylvania, just outside of Pittsburgh, has won multiple awards for their Bamboozle, a beer-washed cheese made with a blend of goat and cow's milk. Because good bloodlines create high-quality dairy goats, many goat farmers devote considerable time and energy toward their breeding program, generating offspring with high-quality dairy genetics that bring a substantial price when sold domestically or exported abroad. Across the board, farmstead goat cheese producers are working harder than ever to develop, nurture, and maintain customer loyalty.

Modern goat dairying. Photo courtesy Dairymaster Inc.

In the modern, ever-changing goat dairy economic landscape, some states have made it a bit easier to make and sell cheese. For example, Maine law allows licensed cheesemakers to heat treat their milk before making cheese, an alternative to pasteurization that ensures food safety but requires less upfront investment in expensive equipment. Cheese produced this way can only be sold within the state of Maine; at the same time, Maine consumers are known to be staunchly loyal to local producers. In 2015, Wyoming passed the Food Freedom Act, allowing the sale of certain categories of foods processed on farms or in otherwise unlicensed facilities. Products created in this manner (with some exceptions) can only be sold to individuals, not stores or restaurants, and only within the boundaries of the state. The law has enabled some small goat dairies to develop a local following for their cheese and other products, including raw milk, without having to foot the bill for the expensive start-up costs of a fully licensed dairy. While most states have similar "cottage food" laws, not all such laws cover dairy products, and the restrictive scope of these laws are specifically designed to preclude the development of a larger scale, profitable business venture. Ultimately, without sufficient capital to meet the many and increasingly expensive demands of farming, including land, and labor costs, let alone dairy plant licensing and product distribution, the small-scale farmstead producer that drove the renaissance of goat cheese production in the United States in the 1980s and '90s has become a quaint anachronism.

Decades after the goat dairy boom of the early twentieth century, a surge in public enthusiasm in the 1970s jumpstarted a revival in the production of goat's milk and cheese. The subsequent rise and rapid expansion of the commercial goat cheese industry in the United States has largely been a success story. Goats dairy products became increasingly associated with health and then gourmet sophistication starting in the 1980s. Goat cheese has since become a staple in the diets of many Americans and is as commonplace in the refrigerated cases at national grocery chains as it is in specialty cheese shops. These days you are more likely to see goat cheese on a frozen pizza than on a gourmet salad served at an expensive high-end restaurant. Goat dairy products have finally achieved the widespread commercial success early industry pioneers dreamed of. In fact, goat dairying and cheese production are entering a new stage of their history, one characterized by international corporate players and commodity-scale industrial production. Where the industry will be fifty years from now is anyone's guess.

III

Contemporary Issues in Goat Culture

CHAPTER SEVEN
Urban Goats

I'm Pro Goat and I Vote!
>Jennie Grant, Goat Justice League, 2007

Like many aspiring urban homesteaders of the early 2000s, Jennie Grant started out keeping chickens in her backyard. The benefits were obvious: she and her family loved the fresh eggs, and Grant appreciated the education her son received about where his food came from. Empowered and energized by the process, her thoughts turned to milk. "I liked to get food from the farmers market, but I couldn't get milk there at the time," she remembers. Meanwhile, Grant and her husband were already in the process of clearing their overgrown backyard, a small space they lovingly termed their back forty. One thing led to another and soon two Mini LaMancha goats, Brownie and Snowflake, joined the chickens in the Grant family's backyard.[1]

Most of the Grants' neighbors found the goats to be a pleasant diversion. The situation took a turn for the worse, however, when a little girl who lived a few blocks away came down with Q fever, a bacterial infection that can be transmitted by livestock. The distraught parents blamed the goats. While it was eventually determined that Brownie and Snowflake were *not* the source of the illness, in the process the Grants were outed as keepers of then-illegal goats within the city limits. The Seattle Department of Planning and Development left an ominous notice on the family's porch declaring that the goats had to go.

At the time, Seattle ordinances permitted farm animals (including goats) only on lots over 20,000 square feet—almost half an acre, a large expanse of land rarely found as a single lot within a densely populated city. Grant initially sought a waiver that would carve out an exception for her well-behaved goats, arguing that miniature goats were not, in fact, farm animals like cows and sheep. Her interpretation of the law was rejected. Undeterred, she wrote to then-Seattle City Council member Richard

Conlin who, along with legislative aide Phyllis Schulman, suggested they work to change the law instead.

As part of the process of introducing a new ordinance before the Seattle City Council, Grant needed to demonstrate community support for her cause—in this case, backyard goats. Ever resourceful, Grant and her son, then six, proceeded to gather signatures, first from neighbors and later at area farmers markets. "But then I realized, you can't just hand someone a clipboard and expect them to sign something," Grant said. "You have to have an organization to sound legitimate." And so the Goat Justice League was born.

Grant's legalization campaign, with the catchy tongue-in-cheek slogan "I'm pro goat and I vote" went viral. Reporters descended on the story and requests for interviews materialized from media outlets all over the country, though Grant assured me, "there were no paparazzi or anything." *Time* called Grant "the Godmother of Goat Lovers." *The New York Times* featured Grant along with urban goat owners from across the country in an article about the urban goat keeping trend.[2]

In September 2007, the Seattle City Council voted unanimously to carve out an exception in its zoning to allow miniature goats. The moment finally brought resolution for Grant, who keeps goats at her Seattle home to this day. The publicity generated through her efforts had a broader effect, inspiring people in a number of cities to work to change their own livestock regulations. Chapters of the Goat Justice League sprang up in San Diego, California; Charlottesville, Virginia; and Lexington, Kentucky. The Goat Justice League also spawned like-minded organizations, such as one called No Goats, No Glory in Colorado Springs, Colorado. Widespread media coverage of Jennie Grant and her goats inaugurated a full-blown urban goat era in the United States.

Grant's high-profile efforts notwithstanding, the not-so-secret secret of the American urban landscape is that goats have always been a feature

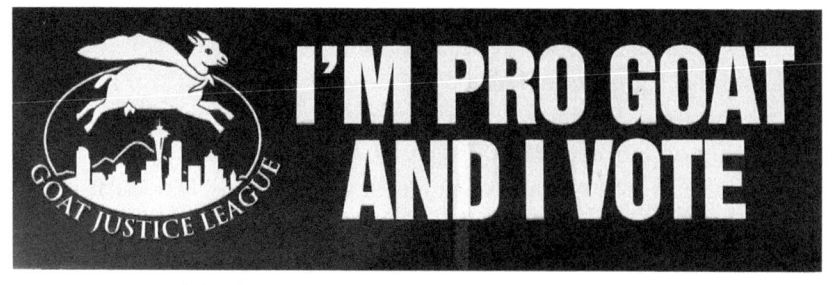

Jennie Grant used the slogan "I'm Pro Goat and I Vote" as part of her campaign to legalize goat keeping in Seattle. Collection of author.

of American urban landscapes. Despite the efforts of citizens and regulators in nineteenth and early twentieth-century cities across the country, goats never really disappeared entirely. The evidence is everywhere: from Fort Worth, Texas, where a goat was the subject of a protracted chase that involved a large part of the night police force (1933) to Spokane, Washington, where the city health officer ordered two goats removed from a city because of their smell (1936), goats have been present in cities and towns across the country throughout the twentieth century—much to the chagrin of legions of city dwellers.[3]

Among the more well-known of the twentieth century's urban goats were a select few residing in Chicago. William Sianis, a Greek immigrant, purchased what was then known as the Lincoln Tavern in Chicago in 1934. A charismatic man, Sianis was a natural at capturing the public's attention, and after a baby goat was said to have fallen off a truck in front of the tavern, Sianis seized the opportunity and turned goats into his brand and calling card. The tavern subsequently became known as the Billy Goat Tavern, and Sianis began to call himself "Billy Goat." Murphy became one in a long line of goats that resided at the tavern and mingled with patrons over the subsequent decades.

You may know how the rest of this story goes. In 1945, Sianis brought Murphy the Goat to the fourth game of the World Series between the Detroit Tigers and the Chicago Cubs at Wrigley Stadium in Chicago. The pair were denied entry to the game, purportedly being told something along the lines of "the goat stays out because he smells." Detroit went on to win the World Series that year. Subsequently, a myth emerged that an outraged Sianis had let loose a vengeful curse on the Cubs, which kept the team from winning the World Series for many decades. In short, Chicago Cubs fans (supposedly) had an urban goat to thank for their team's decades-long championship drought. Thankfully, the Cubs finally won the World Series in 2016, presumably breaking the curse for good.[4]

On the West Coast, urban goat owner Estelle West came to the attention of San Francisco officials in the 1950s. West lived with her goats in the Potrero Hill neighborhood; authorities appear to have tolerated her goats despite the city's two-goat limit because West lived in what was then a relatively remote section of the city south of downtown populated mostly by the working class and immigrants. Ignorance turned to intolerance when it happened that West's property was in the way of the state highway department's plans for constructing an extension to the Bayshore Freeway (formally known as US Highway 101). A protracted fight ensued; at one point West dramatically fed the sheriff's papers to one of her goats.

"Woman and her Eighteen Goats Defy Law and Progress," screamed the headline in the *San Francisco Examiner*. West eventually surrendered her property, receiving $3,500 for her troubles. But the story was not over; just a few years later the so-called Goat Lady of Potrero Hill reappeared in newspaper headlines when she refused to get rid of her goats after neighbors complained repeatedly of the smell. This time a judge actually sent West to jail, but in an only-in-San Francisco moment reported in excruciating detail by the *Examiner*, she was bailed out by a local burlesque dancer turned socialite.[5]

Urban goat antics persisted throughout the twentieth century. In 1971, a billy (male) goat was on the loose in Manheim Township, just north of Lancaster, Pennsylvania, "generally making a nuisance of itself by browsing on well-manicured suburban lawns and napping under expensive shrubbery." In 1979, the city council of Albany, California, reversed a long-standing ban on backyard goats after a city resident applied for permission to keep two nanny goats in his backyard. Neighbors rallied behind the cause and packed the city council hearings. According to reports, the plucky applicant passed around a jar of goat droppings to the council members to prove that goat droppings don't smell. The council members' reactions to the jar of goat poop were said to have ranged "from surprise to guarded approval." Authorities in Lincoln, Nebraska, were less impressed by the potential of backyard goats. In 1995, they forced a fourteen-year-old girl to give up her Pygmy goat, Dorothy, because goats were not allowed within the city's limits. "The city is not exactly a farmstead situation," said the city's animal control manager, who gave the girl ten days to comply.[6]

Negotiating the Presence of Goats in Urban Spaces

It's no exaggeration to say that goats have been a constant, if not always welcome, presence in United States cities for as long as there have been cities. Although the reputation of urban goats has not always been positive, it's clear that the cultural conversation about goats has shifted perceptibly. Increasingly, municipal officials and urban planners are focusing on ways to responsibly include goats, as well as chickens and other small livestock, in urban environments. To date, many of the top fifty cities in the country by population allow goats and other small livestock, and hundreds of smaller cities have legalized goat keeping within their jurisdictions as well.

Many municipalities employ land-use regulations to regulate small livestock, though the scope and character of those laws vary considerably. One of the most common types of regulations are those concerning the spaces where animals are kept. These laws proscribe designated lot sizes

Estelle West kept goats in the Potrero Hill neighborhood of San Francisco. During the 1950s, authorities forced West to give up her goats and move because freeway construction was set to run directly through her property. San Francisco History Center, San Francisco Public Library.

and setback requirements. For example, in Fayetteville, Arkansas, goats may be kept within the city limits only if the lot is 10,000 square feet or larger, along with an additional a twenty-five-foot setback between the goats and any neighboring residence. Similarly, if you want to keep goats within the city limits of Des Moines, Iowa, setback requirements designate that your livestock shelter must be "35 feet from a property line and 45 feet from any adjacent dwelling unit. Roaming or grazing areas must be at least 20 feet from any property line and at least 30 feet from any dwelling unit." Cities such as Baltimore, Maryland, employ a more elaborate step system. You may keep up to two goats, which must be licensed, on any residential property in Baltimore; for lots greater than 20,000 square feet you can add an additional goat for every additional 5,000 square feet, up to a maximum of six goats.[7]

Some municipal codes impose specific requirements for livestock keeping practices. Such site-level rules may include enclosure parameters and manure disposal plans. In San Diego, for example, goat keeping requirements include the presence of a predator proof shed of at least ten square feet surrounded by a fence that is least five feet tall, all easily accessible for cleaning. In Columbus, Ohio, prospective goat keepers must undergo an application and review process and submit detailed plans for how the resident goat(s) will be housed, along with additional information concerning sanitation practices and care and health procedures. Applicants must also allow city's public health veterinarian to inspect their premises.

If there is an urban livestock utopia in the United States, it may be in Compton, California. At first glance, the Richland Farms neighborhood in Compton looks like any other Southern California neighborhood, with stucco houses and palm trees as far as the eye can see—that is, until you see riders on horseback ambling along the side of the street. Richland Farms is something of a unicorn within the city of Compton, and within the greater Los Angeles area, with large residential lots of one acre or more; many lots have been combined over the years, making their footprint event larger. The neighborhood teems with all manner of livestock, including horses, cattle, chickens, rabbits, and goats. Several other neighborhoods in the region retain a similarly agricultural character, including Melody Acres in Tarzana and Walnut Acres in Woodland Hills, both in the San Fernando Valley.

Legend has it that the Compton neighborhood's namesake, Reverend Griffith Compton, stipulated that the Richland Farms neighborhood be preserved in perpetuity as an agricultural oasis, but the roots of the neighborhood's agricultural character are more likely rooted in the broader development of the Los Angeles area during the early twentieth century. While San Francisco was the largest city in the state for many decades after the Gold Rush, during the early twentieth century the population of Los Angeles skyrocketed, going from just over 100,000 in 1900 to over 1.2 million by 1930, quickly taking over as the state's most populous city. During this period of rapid expansion, real estate developers enticed buyers to travel to the Los Angeles region by offering "small farm homes" set on lots ranging from one to three acres. Prospective residents were enticed by the idea that they could move to California and live self-sufficiently by growing and selling produce and other farm products. Like other neighborhoods across the region, Richland Farms was positioned as a potential agricultural oasis; ads promoted its "Income Producing One Acre Homesites" and "Splendid Cooperative Water system" located right in the neighborhood. These and other incentives worked: during the first half of the twentieth century Los Angeles County was the top producing agricultural county in the nation.[8]

During its earliest years, the city of Compton attracted white buyers seeking California's advertised agricultural Eden. Over the decades, however, Los Angeles's farming culture faded, and the city once known for its agricultural productivity slowly transformed into the vast urban sprawl that it is today. Black families began to move to Compton starting in the 1950s and '60s as discriminatory real estate practices redlined them out of many areas of Los Angeles. Now Richland Farms consists of a mix of Hispanic and Black residents, a number of whom keep goats for both milk and meat. And if you know where to go, you might find a goat farmer serving *pajarete*, a Mexican specialty drink made with raw goat's milk. The pajarete tradition originates in the Mexican state of Jalisco and the surrounding region; there, ranchers blend raw milk straight from the animals (goats or cows) with chocolate, coffee, and a dose of alcohol to create a bracing beverage to start the day. In Compton, pajarete delivers a sense of community and tradition.[9]

❖

Community discussions about urban goats often begin when a neighbor complains about goats in someone's yard. That's what happened to Jennifer Council, who keeps two goats, Mirage and Diddy, in her backyard in Brighton, Colorado, a Denver suburb with about 40,000 residents. An avid gardener, Council produces a prodigious amount of food in her backyard, including a variety of vegetables; she also raises chickens and rabbits along with the goats. Council came home one day to find a notice on her door saying she had to get rid of the goats.

The Brighton City Council then took up the matter of urban goat keeping; among its first acts was to conduct a poll to assess residents' opinions about goats within the city limits. The poll revealed common concerns often seen in discussions about urban livestock, including worries about potential smell and noise. A number of residents also expressed concern about the city's ability to handle the logistics of animal keeping within the city limits, but overall Brighton residents supported keeping goats by a small margin. Discussions continued for almost a year; in the meantime, Council boarded her goats at a farm outside of town.

Eventually the Brighton City Council authorized a two-year pilot program allowing goats within the city limits. The pilot program came with a laundry list of requirements: residents interested in keeping goats were required to enter a lottery for just ten pilot permit slots. Any goats would have to be miniature female goats no more than twenty-four and a half inches tall, vaccinated for rabies. Additional rules specified property lot size and shelter requirements. In a unique twist, the city council also required applicants to pass an open book test covering goat keeping issues, and test takers were required to score at least 80 percent on the test in order to move forward.

In the end, nine people entered the Brighton goat lottery. Based on the success of the pilot program, after one year the Brighton City Council voted unanimously to remove all restrictions on goats within the city limits in March 2023. A city councilor who had initially opposed legalizing goat keeping remarked, "It hasn't turned out to be the fiasco everyone imagined . . . I am very, very happy that we are going to make [legal goat keeping] permanent." While the process caused Jennifer Council quite a bit of stress, in the end she was able to bring her two goats, Mirage and Diddy, back home.[10]

Laws regulating urban goats vary widely across the country. While it is legal to keep goat in a number of large cities, including Chicago, Seattle, and Los Angeles, other cities have not gone that far. This kind of ban is especially puzzling in Detroit, a city which has a long history of urban agriculture. In the 1890s, Detroit Mayor Hazen Pingree, responding to a national economic recession, allowed citizens to appropriate vacant city lots in order to grow food. The Pingree Potato Patch Program is considered one of the earliest examples of organized urban agriculture in the United States. Despite this, for many years Detroit authorities have taken a particularly strong stance against urban livestock, going so far as to seize three goats and six chickens from two distraught residents in 2014. Even after that well-publicized incident, authorities continued to issue citations for keeping small livestock. More recently, signs of progress have emerged. In early 2024, Detroit City Council President Pro Tem James Tate introduced a proposed ordinance that would allow city residents to keep bees, chickens, and ducks within the city limits with a city license. The proposed ordinance would not permit residents to keep goats or other animals, including rabbits, however.[11]

One of the main areas of concern often expressed by municipal officials with regard to urban animal keeping is animal welfare. In 2019, Chicago Alderman Raymond Lopez suggested that he might introduce an ordinance strictly regulating livestock of all kinds, including chickens and goats. Chicago currently takes a particularly lenient stance on livestock of all kinds within city limits, with the only restrictions being those covered by noise and nuisance laws. The alderman, known for his animal welfare stance, voiced concerns about large cockfighting rings, which have long been a problem within the city. His proposed measure would have required livestock owners to obtain permits and limit the numbers of permitted animals allowed to be kept. The alderman's campaign took on further momentum after animal control officers found a dead horse in the Englewood neighborhood. Despite his efforts, momentum for making changes soon fizzled. One Chicago resident told me that the alderman "didn't realize who he was dealing with. The backyard chicken people in this city have money and will fight."[12]

The enduring presence of animals in cities, regardless of whether or not they are legal, suggests the potential of other types of activities, including the slaughter and use of animals for food. As urban agriculture and animal keeping becomes more widespread and adherents grow more serious about raising their own food, the issue of regulating the scope of acceptable practices has come to the fore. Among regulators' concerns with these practices are the potential spread of salmonella or other pathogens, as well as the illegal sale or distribution of home-processed meats.

Urban slaughter and animal processing became a particular source of contention in Oakland, California. There, urban farmer and activist Novella Carpenter, author of *Farm City: The Education of an Urban Farmer*, clashed with city officials in 2011 when she sold rabbit pot pies using meat from rabbits she raised on a small plot of land next to her home. As a result of the efforts of Carpenter and others, Oakland city planners set out to overhaul the city's urban agriculture code. The process led to protracted debates over the ethics of animal keeping and the best way to regulate or allow associated activities such as animal slaughter and sales. An organization called Neighbors Opposed to Backyard Slaughter formed to oppose proposals to allow urban animals. "It is wrong to believe that the backyard keeping of animals is anything but a threat to the health and wellbeing of our community," wrote the organization's co-founder in an op-ed piece. According to one report, three hundred people attended an Oakland Planning Commission's public meeting where debate over the slaughter issue was heated. The public outcry delayed revisions to the city code by several years. As of this writing, it is currently legal to keep goats and other small livestock in Oakland, though the slaughter of any kind of animal is prohibited within the city limits.[13]

Disagreements over urban animal keeping and related activities are often rooted in cultural differences. In Miami, Florida, citizens are used to finding a variety of animal parts, including cow tongues and goat heads, strewn in parks and other areas around the city. The remains are left by practitioners of Santeria, an Afro-Cuban religion practiced by many in South Florida; the practice is part of a ritual of feeding or otherwise appeasing the religion's deities, known as orishas. Over the years, citizens not affiliated with the religion have often expressed repulsion over such practices. The issue was actually resolved several decades ago when the Church of the Lukumi Babalu Aye, a Santeria church, was established in the city of Hialeah, Florida, a Miami suburb. After the city of Hialeah passed several ordinances prohibiting ritual animal slaughter, the church sued. The issue eventually came before the US Supreme Court, which ruled in *Church of the Lukumi Babalu Aye, Inc. vs. City of Hialeah* in 1993 that Hialeah's attempts to regulate animal slaughter violated Santeria practitioners' rights to freely exercise their religion. Although the issue was settled in a legal sense, conflicts between Santeria practitioners and other residents continue in South Florida. As one Miami resident said more recently, "I am respectful of any religion, but not when you are performing your rituals [such as leaving animal parts in city parks] in a public place."[14]

The Urban Goat Experience

Given the sometimes-complex web of bureaucratic requirements that can accompany keeping goats in cities, it's worth asking—why would someone bother keeping goats in cities in the first place? Among the dozens of urban goat owners I spoke with, I found a wide range of motivations. Many simply want to keep goats as pets. "I had a potbellied pig growing up and that's what started my love affair with farm animals," an urban goat keeper in Atlanta, Georgia, told me. She keeps two Nigerian Dwarf goats in her backyard in the Buckhead neighborhood. Goats are friendly, curious, and engaging animals—many people compared them to dogs—so this makes a lot of sense. Another goat owner in Portland, Oregon, told me she first got goats because she wanted to get her kids into hiking, and she hoped that the goats could serve as pack animals while the family was on the trail. A mom in Phoenix, Arizona, said that her family first got goats after keeping chickens for a few years, because their daughter was lactose intolerant, and they wanted a goat for its milk. Many urban goat owners spoke of a gateway animal, usually chickens. Once they'd acquired chickens—or bees, another common gateway animal—they became enamored with the experience of producing their own food. For some, goats were the next logical step.

Carolyn and David Ioder purchased two goats in 2011 for their property in the South Austin neighborhood of Chicago. GlennArt Farm (named after Carolyn and David's fathers) kept around eight to ten goats, most of them Saanen-Toggenburg crosses. While the goats lived mostly in the family's backyard and converted garage, during the spring and summer months, the goats were herded a short distance down the alley behind their house to a nearby community garden, which allowed the goats to graze on part of the open lot. The alley trek—an unexpected moment of urban transhumance—is quite a scene. One morning I assisted the Ioders, herding stragglers from the back of the pack, as firefighters from the firehouse next to the community garden watched the organized chaos of goat herding unfolded before them. In 2018, GlennArt Farm became a licensed dairy in the state of Illinois; Carolyn called it "the smallest dairy in the world," and that's probably not far from the truth. For a number of years the farm sold raw goat milk and held regular goat yoga classes as well as, other goat-themed events. More recently GlennArt has closed their milk business, though they continue to keep several goats and conduct goat-related activities like what they call "goat chills," where attendees can come to the farm and hang out with the goats.

While many people successfully and happily keep goats in urban spaces all over the country, it's worth noting that the practice presents unique challenges. Although goats can be cute and loveable, their behaviors and needs are very different from the cats or dogs that animals lovers may be more used to. Gianaclis Caldwell, longtime goat owner and author of several books, including *Holistic Goat Care*, said that "goats can do just fine in urban spaces. But owners should be careful to educate themselves about the needs of the animal. Goats don't necessarily need pasture but they do need hay, and they love tree branches and other plant materials," she said.

Carolyn Ioder of GlennArt Farm milking one of her goats in Chicago's South Austin neighborhood. Author photo.

As for choosing a specific animal, Caldwell advises that "unless the goat is going to be used for milk, I'd recommend getting a wether (castrated male goat). These goats can be ideal for urban goat keeping because they will not go into heat like female goats do—which can be quite noisy." Most urban goat owners I talked to recommended that newbies take the time to talk to experienced urban goat owners before taking steps toward actually acquiring goats.[15]

In addition, the expenses of urban goat keeping can also be significant; buying and hauling hay and other types of animal feed can be both a financial and logistical hurdle. Livestock feed stores are not typically found in cities, though more recently the urban farm store phenomenon has sprung up to fill that niche. In addition, there is the issue of access to and cost of health care; veterinarians with small ruminant expertise are typically not found at an average city veterinary practice—in fact, there is a currently a national shortage of veterinarians with small ruminant experience, meaning that even rural goat owners may have trouble finding veterinary care. Several urban goat owners I talked to travel long distances to rural veterinarians or to university veterinary clinics to access care for their animals. In an emergency there may be no medical assistance available at all.

Urban spaces present a number of other types of challenges to goat owners. Goats are naturally curious and outgoing and nearly every city goat owner I talked to had a story about the goats escaping from a backyard enclosure. City goats also tend to attract interest from passersby who are sometimes inclined to try to feed the goats. While usually well meaning, this kind of interest can turn harmful if the goats are fed foods that are not goat-friendly—despite the lingering stereotype that goats will eat anything, they actually have very sensitive digestive tracts. One clever urban goat owner tackled this problem by repurposing an old candy vending machine and placed it outside the goat pen near the sidewalk, and visitors could purchase a handful of proper goat food for twenty-five cents. Despite the occasional challenges, the majority of goat owners I spoke with said that their experiences with the public are overwhelmingly positive. People stop by regularly and children are awed by the creatures.

Occasionally, the public's attention is not so positive. "I have had random people show up in my backyard, wanting to bring their children and grandchildren," said one goat owner. Another told me she had to put up a no-trespassing sign to remind people that her yard was not a public access zone. Dog attacks are another fairly common hazard; city dogs unfamiliar with larger animals like goats sometimes act unpredictably, as do stray dogs. In extreme cases, the attention can turn destructive. Several years ago in Portland, Oregon, a goat owner woke to find that one of his three Nigerian Dwarf goats had been kidnapped from his backyard. Within a couple of hours the goat was found in a park ten miles away with a broken leg and severe lacerations. Injured but mostly OK, the goat wore a cast and underwent several months of recovery and physical therapy for her injuries. The goat now roams happily with her two goat companions; the owner has since installed security cameras around his property. The moral is that vigilance is one of the many keys to successful urban goat ownership.

Those aspiring to keep goats should understand the many challenges that come with the practice; that being said, the joys of owning goats are their own reward.

❖

Once reviled and feared, urban goats don't have time for us mortals anymore—they are too busy promoting themselves on their social media feeds. Social media has opened up a new means by which city dwellers (or anyone really) can commune with goats, with none of the hassles or responsibilities of actual goat ownership. *INStyle*, a popular beauty and fashion magazine and web platform, once declared: "It's Official: Goats are the New Puppies of Instagram."

The many and ever-expanding forms of social media have enabled goats to infiltrate hearts and minds in an infinite variety of interesting ways. One of the early moments in the ascension of goats to global social media dominance was a video titled "Goats Yelling Like Humans," first posted in 2013. The video spawned what has become known as the screaming goat meme, which consists of screaming goat sounds placed in music videos at strategic points, with the results then passed around social media. Another popular goat-themed video, "Goat Babies in Pajamas," was first shared by Sunflower Farm Creamery in Cumberland, Maine, in 2015. Owner Hope Hall initially posted the video for fun and didn't think anything of it. Since then, the endearing video of baby goats cavorting while wearing colorful pajamas has received over ten million views (and counting), and its widespread popularity transformed their small farm business into an agritourism destination. "People started visiting us from all over the world," Hall said. "We hope that the farm experience helps them understand food and agriculture in a different way." A few years later, during the Covid-19 pandemic, actor Kevin Bacon introduced his considerable social media following to the two goats he bought for his wife, Kyra Sedgewick, on their wedding anniversary. One day Bacon began singing to the goats, Louie and Macon, as a way to relax. The sessions were memorialized with their own hashtag, #GoatSongs, and Bacon has since covered songs by a variety of artists.

And then there is goat yoga. While goat yoga is now a popular worldwide phenomenon, the origins of the concept trace back to rural Monroe, Oregon. Founder Lainey Morse was going through some difficult times in 2016 when she had the idea of having friends over to her small farm occasionally for what she called a "goat happy hour." Friends stopped by,

Images of goat kids wearing sweaters or pajamas, like Otis pictured here, have become social media staples. Photo courtesy of Sunflower Farm Creamery.

kids in tow, and everyone would have fun hanging out with Morse and her goats. One day, one of the moms in attendance, yoga instructor Heather Davis, asked if she might hold a yoga session at the farm and harness some of the peace and serenity. Goat yoga was born. Morse's marketing background came in handy when the idea caught on after her article about the first goat yoga session appeared in *Modern Farmer*. Like the Sunflower Farm Creamery video, goat yoga quickly went viral. "It was crazy, I was getting calls all day, every day," says Morse. Since that first quiet afternoon on an Oregon farm, goat yoga has expanded across the globe. Goat yoga classes are popular in many countries, including Germany, France, and the Netherlands.

In addition to introducing the practice of yoga to a whole new generation of enthusiasts, goat yoga helped popularize goat-themed events of all kinds. These days you can rent goats for just about any occasion; businesses across the country will bring goats to birthday parties, weddings, or other events and celebrations. In a similar vein, vacation rental services now offer goat-centered stays and goat-themed experiences like goat hikes. The Covid-19 pandemic spawned another concept: goats attending online meetings, an idea that livened up the work-from-home doldrums, not to mention boosting goat owners' incomes. "During the pandemic we would walk out into our pasture with a camera and give people an online tour of the farm," said Margaret Hathaway of Ten Apple Farm in Gray, Maine. "I think it really helped folks remember the natural world was still out there."[16]

A scene from the first ever goat yoga session held in 2016. Photo courtesy of Lainey Morse.

❖

One sunny February afternoon, I went for a walk with Erica Somes and her two goats, Clayton, and New York. Erica is an enthusiastic urban farmer in Portland, Oregon, who keeps goats, a potbellied pig, and chickens. She is well-known in her neighborhood for her regular goat walks. Clayton and New York are two handsome Nigerian Dwarf goats, tawny brown with striking black markings, about the size of a largish dog, although the dog would not be sporting Clayton and New York's gorgeous horns, which lend a distinct Nordic warrior vibe.

Erica is an expert goat walker, and both of the boys are well trained. They clearly adore her and follow her every move and command, whether she tells them to slow down, keep going, or stop eating the bushes. In fact,

Clayton and New York already know which neighbor's bushes are OK to munch, and usually head for those yards.

Within a few minutes of setting out on our walk, a forest green minivan pulled up beside us. The driver's window rolled down slowly. A beaming woman took out her phone and recorded the scene. "This is why I moved to this neighborhood!" she laughed, smiling broadly. "Everyone is so friendly!" This happened repeatedly throughout our walk—pedestrians smiled and took photos.

Despite criticisms from the goat skeptical, one of the positives of modern urban goat is that it creates community. Goats draw people in for conversations and interactions; suddenly neighbors have something in common. "My neighbors all know me," Erica says, "and they love the goats." An urban goat owner in Pittsburgh, Pennsylvania, once received a note that read in part, "I discovered your goats while I was waiting for a bus on the corner and heard them bleating . . . I couldn't resist going between the houses to see them . . . They're just so darn cute to visit on these beautiful fall afternoons." How many of us can say that we know all of our neighbors, or that we have something to talk about with them other than the weather or the occasional downed branch in the street? Goats bring people together.

In Altadena, California, just north of Los Angeles, urban goat owners Choi Chatterjee and Omer Sayeed started out on a personal mission toward achieving a more sustainable footprint. The couple first acquired chickens, then two Pygmy goats, Daisy, and Blueberry, to provide compost for their expanding garden. The productive gardens soon spread out of their yard and onto their median/parking strip, where they planted fruit trees, herbs to attract butterflies, and sweet potatoes, among other things (it is Southern California, after all), transforming what is typically an unused area into what Chatterjee calls a "multipurpose community space." Neighbors pick the fruit and herbs, and many have become friends. "Some neighbors have been inspired to start growing their own vegetables as well." Education is an important component of their vision: Chatterjee and Sayeed offer classes to help teach others to learn the skills they've acquired, covering such topics as compost management, vegetable gardening, raising chickens and goats, and cooking with fresh garden ingredients. "At some point the market took over the functions of the family," Chatterjee says. "We think of food production as a way of being, rather than a state of passive consumption."

In a number of goat-friendly cities, nonprofit organizations are further amplifying the role of goats as charismatic community ambassadors.

Among these is the Philly Goat Project, started by Karen Krivit and her daughter Lily Sage in Philadelphia, Pennsylvania, in 2018. Among the organization's wide variety of programs are weekly goat walks on city streets. "When we walk by people we invite them to walk with us, we always stop and let people take pictures. We tell them who the goats are named after and why they're named after that person. We try to engage people in every possible way," said Krivit. Philly Goat Project also employs people in the community to take care of the goats. A former social worker, Krivit sees lots of opportunity in working with students in the community, especially special needs kids. "Recently we worked with a school on Philadelphia where all of the kids have some kind of physical disability, and were either walking [with] walkers, or in wheelchairs. This school typically has difficulty with family engagement, but when we were there with the goats everybody wanted their picture with their kids and the goats!"

In San Francisco, the goats of City Grazing are often spotted around town, munching contentedly on the city's weed-strewn hillsides. The nonprofit focuses on sustainable land management through goat grazing, and the goats also attract lots of attention. "When you've got a group of goats eating stuff down, dampening fire hazards, it's inspiring to people," says Genevieve Church, the group's executive director and chief goat wrangler. "It's a way to show kids nature in action, helping nature. Goats are also a great way for parents to connect with their kids, and for communities to come together." City Grazing's regular appearances in brushy spots on the Presidio and at the University of California, San Francisco are so popular that they've become events. "People will call us to ask when it's coming up, and plan around it."[17]

❖

As of this writing, it's been nearly two decades since Jennie Grant's campaign to legalize goats in Seattle. At that time, Seattle City Council member Richard Conlin predicted that "possibly hundreds" of goats would eventually be legalized in Seattle once the laws were loosened. Despite Conlin's optimism, Seattle recorded just thirty-one licenses for goats in 2019.

There's no authoritative source that tracks the numbers of goats living within cities across America. The numbers are a moving target; some cities and towns require licenses, which can lead to a more precise count, but others do not. As a result it is difficult to definitively quantify the true scope of urban goat keeping in the United States. And there will always be

Goats from City Grazing at work on a hillside in San Francisco. Author photo.

a certain number of goats kept illegally within cities, no matter what the rules. But what I can do is offer some numbers that help convey sense of the recent state of the practice: An academic survey of 140 urban livestock owners across the nation conducted in 2014 turned up just seven owners of goats, versus ninety owners of chickens. Between 2016 and 2020, three goat permits were issued in Columbus, Ohio. In Baltimore, twelve goat permits were issued between 2012 and 2021; permits are not updated after issuance, so those numbers do not necessarily represent a current number of goats in the city. While it's hard to draw conclusions from such disparate numbers, one thing is crystal clear: while interest in urban goats definitely exists and is ongoing in cities across the country, fears that urban goats will take over are greatly exaggerated.[18]

For centuries, citizens have attempted to keep animals out of villages, towns, and cities across the United States in an effort to rid the populace of many problems, including smells and potentially destructive or dangerous behavior. In the twenty-first century, minds are changing, and the role of goats in contemporary society has grown to encompass social welfare, community building, and land management. Now many cities and citizens are welcoming animals of all kinds, including goats, into their midst. Ultimately, the continued presence, and staying power, of goats in urban spaces demonstrates that the central issue is really not legalization as such, but how cities and urban residents will continue to adapt to the ongoing presence and increasing popularity of goats.

CHAPTER EIGHT

Goat Meat in America

Goats are noble creatures of great utility, and it's time someone put some work into their PR.

James Whetlor, 2018

In 1922, the Texas Sheep and Goat Raisers Association held a contest. The livestock and ranching trade group was determined to select a name for a product important to the group's members—Angora goat meat. At the opening ceremony of the organization's annual meeting that year, a winner was announced: the new and improved designation was "chevon," a fanciful combination of two French words, *chèvre*, the word for goat, and *mouton*, for sheep. One of 2,500 contest entries submitted from all over the country, the winning term was submitted by Mrs. E. W. Hardgrave, wife of a Texas goat rancher. Mrs. Hardgrave was awarded a registered Angora buck for her winning entry.[1]

Although seemingly innocuous, the ceremonial selection of a new name for goat meat was more than an entertaining party diversion. The exercise was motivated by several decades of public controversy, and the stakes were high. During the early twentieth century what might be called a "goat meat panic" was spreading across the country as the public became aware of the fact that butchers and meatpackers customarily marketed goat meat as mutton. The problem, at least in the eyes of many consumers, was that the term "mutton" was commonly understood to refer to meat from sheep. Consumers did not perceive sheep and goat meat as equivalent products, and the apparent intentionality of the deception only made matters worse. Media outlets picked up the story. *The New York Times* railed "Goat Meat Sold for Mutton," reporting that goats were being consigned for sale at livestock auctions, butchered, and deceptively marketed as mutton in Kansas City. A Pittsburgh newspaper headline warned that "Housewives Had Better Keep an Eye on Their Butchers," since a train carload of goats had just arrived in the city from Chicago and "a large number of

Pittsburghers will be buying goat meat instead of mutton in a day or so." Public health officials in Pennsylvania devoted particular attention to the issue, going so far as to conduct a sting operation in Philadelphia during which several storekeepers were arrested and fined for selling goat meat labeled as mutton. As newspapers fueled the flames of public controversy, it's not hard to understand why Angora goat ranchers were worried.[2]

❖

While goats are among many species of domesticated animals that have inhabited the United States for centuries, the Angora breed first arrived on the continent during the mid-nineteenth century. Dr. James B. Davis, a South Carolina physician turned farmer, traveled to Turkey at the behest of President James Polk during the 1840s to educate the population about cotton production. Davis returned to the United States in 1849 accompanied by a variety of animals from the Mediterranean region, including Brahmin cattle, water oxen, and Angora goats. Angora goats are a species native to Turkey, and the name is an anglicization of Ankara, Turkey's capital. The goats are cultivated for their long curly coat; angora wool, commonly known as mohair, has been used for weaving and textile production around the world for centuries. Dr. Davis was said to have imported the animals in order to develop a mohair trade that could jumpstart the depressed economy in the heavily cotton dependent American South.[3]

Despite Davis's enthusiasm, mohair production never took off in the South because the hot, humid climate proved ill-suited to raising Angoras. But agricultural entrepreneurs seized on the idea and began importing additional animals from Turkey and South Africa. Within a few decades Angora ranching began to spread across the country; by the 1880s there were large populations of Angoras along the West Coast in California, Oregon, and Washington. Angora ranching became especially popular in Texas, where wool production boomed after the end of the Civil War. A survey of the sheep industry by the US Bureau of Animal Industry (BAI) noted that there were over 5.1 million sheep in the state in 1890, producing thirty million pounds of wool per year. That same year, the Angora goat population in Texas had grown to 275,000 (though population estimates vary widely). The BAI advocated for continued expansion of Angora populations: "It is very much to be hoped that goat husbandry will be studied by the stockmen of Texas, so that . . . our own markets may be supplied with mohair from our own flocks [rather than imported from outside the

Prize winning Angora goats from Oregon, 1905. Photo by George M. Weister, Angelus Studio Photographs, 1880s–1940s, PH037_b065_LS00088, University of Oregon Libraries.

country]." By the turn of the twentieth century, Texas ranchers produced by far the most mohair of any state.[4]

As the goat meat labeling debate began to spread, the meat industry denied charges of deceptive labeling. It's also worth noting that at the time, no federal laws specifically prevented such practices. The trade publication *Chicago Livestock World* called out the "oft-exploited fallacy that immense quantities of goat meat [go] into consumption disguised as mutton," claiming that goats hardly ever made it to the slaughterhouse floor—notably not specifically denying the practice of substituting goat meat for mutton. New York City butchers dodged the issue by claiming that goat meat was mostly consumed by the "poorer classes. . . . [t]he ordinary customer of cheap mutton cannot . . . cannot distinguish goats from mutton." The controversy became heated enough that D. E. Salmon, head of the BAI, weighed in on the debate: Salmon pronounced that he did not anticipate much danger from goat meat consumption, as goats were generally healthy animals—a statement that surely did little to soothe the nerves of wary consumers. Others attempted to emphasize the potential virtues of goat meat. One Iowa rancher hoped that "the prejudice against goat meat will soon disappear," arguing that goat meat was actually of higher quality because "the animal is a ruminant and eats leaves and bushes and twigs, the most succulent part of the growing vegetation." Ultimately, the lingering association between goat meat and deceptive practices had the effect of focusing the public's attention on a perceived fraud.[5]

It was not long before the outcry over deceptive meat labeling practices drew the attention of the nation's politicians. In 1919 and again in 1921, Arizona Representative Carl Hayden introduced a bill in the US House of Representatives that would have required goat meat to be specifically labeled *goat meat*, though his efforts did not get past the Committee on Agriculture. Since federal laws of the period only regulated meat sold in interstate commerce, a number of states took the opportunity to pass their own meat-specific labeling laws. Missouri and Oregon, among others, passed laws requiring butchers operating within their borders to label specialty meats like horse, mule, and goat meat specifically as such when offered for sale. Oregon legislators later went a step farther, entirely banning the sale of billy goats for meat consumption because their strong odor was thought to cause the meat to taste bad.[6]

The practice of calling goat meat "mutton" may not have been an intentionally deceptive practice, at least not initially. In the Southwest, sheep and goats were kept and grazed together and often referred to collectively as sheep by shepherds and ranchers. This may have followed the practice of Spanish colonists who originally introduced goats and sheep to the Americas: the Spanish typically referred to goats and sheep collectively as *ganado menor* (small livestock). There is some evidence that the term "mutton" was used in the Southwest as a generic term for meat from *any* small ruminant: the term "mutton goats" appears fairly regularly in Texas and New Mexico newspapers during the late nineteenth century. In 1887, one Texas newspaper referred to a rancher driving "mutton goats" from Texas to Nebraska, a journey he had taken several years earlier with his "mutton sheep." The BAI refers to "Angora Mutton" multiple times in its 1900 Annual Report. And in some parts of the Caribbean and Southeast Asia, "mutton" continues to be used to refer to both sheep and goat meat.[7]

Discussions about the proper terminology for goat meat began in earnest when Angora ranchers began to send large quantities of goat meat into the national marketplace. As Angora ranching was beginning to grow in Texas and the Southwest, the cattle-ranching business was rapidly expanding. A vast integrated transportation and distribution network began to develop during the late nineteenth century, which created a streamlined system of meat production and distribution in the United States. Cattle raised in Western states where land was cheap and plentiful were shipped via railroad to finishing centers, often Kansas City, Missouri, after which the fattened animals made their way to slaughterhouses, primarily in Chicago. Families with now-familiar names like Armour, and Oscar Mayer, made vast fortunes in the meat industry. In simpler times, consumers ate meat from animals they owned or perhaps purchased meat from a nearby farm. By the nineteenth century, the meat served on American tables in Boston probably originated in Texas, was transported to Kansas City, and was slaughtered and packed in Chicago. After that, the carcass was shipped to Boston, where it was broken down by local butchers, sold to restaurants or individuals, and consumed. As the population of Angora goats began to increase around this same period, Angora ranchers hoped to capitalize on their animals for meat in the same manner. Whether or not they realized it at the time, by doing so Angora ranchers would find themselves answering to the tastes and scrutiny of the American public, many of whom were not familiar with goat meat. The results were disastrous for the industry.

❖

William L. Black was a St. Louis stockbroker who purchased 32,000 acres of land in central Texas in a region known as the Edwards Plateau in 1876. Black started out raising cattle and sheep on his new ranch and later purchased a flock of goats to supply the Mexican herders he had hired who preferred to eat goat meat. After learning about the growing population of Angora goats in the region, Black became convinced of the breed's potential. Black eventually acquired twelve purebred Angoras, eight males and four females, in order to "upgrade my common goats."

Within a few years the goat population on Black's ranch had multiplied to 8,000 goats. Subsequently Black contacted an acquaintance at Armour and Company, a Chicago meatpacking outfit, about the possibility of selling the goats for meat. The acquaintance replied that "goat meat was not at all fashionable," and he would not be able to purchase any of Black's goats "until the people overcome their prejudice for [goat meat]." Black's contact suggested that the rancher process the meat and sell it in cans himself. In response, Black constructed his own meat canning plant, the Range Canning Company, as well as the adjacent Fort McKavett Tanning Company for hide processing, essentially creating a self-contained goat processing industry on his ranch.[8]

Black was just one of many Texas Angora ranchers trying get the nation's consumers interested in goats and their meat. Despite the controversies, Angora ranchers across the country were working hard to develop a market for goat meat, an additional source of income that could be derived from their herds. W. E. Farlowe, a livestock agent for several railroad companies, was scouting livestock in Texas during the late nineteenth century when he became convinced of goat meat's potential, maybe because a persuasive goat rancher had lobbied him about its value. Farlowe subsequently shipped several train cars of Angora goats to Chicago, hoping to sell the animals to the city's butchers. After a few days of solid rejections, Farlowe had an idea; he held a formal dinner and invited the butchers to try different preparations of goat meat, "roasted, boiled, stewed and fried." Within a day, Farlowe was reported to have sold all of his goats. Angora rancher A. B. Hulit of Missouri, a partner in the Frisco Livestock Company, tried a similar sales approach, organizing a well-publicized goat meat banquet in St. Louis in 1902. According to Hulit, "the purpose of t[his] banquet is to practically demonstrate that goat meat is quite as tender and palatable as mutton." Among the delicacies served at the event were Angora broth, goat chops, braised leg of Angora with walnut sauce, and goat's milk cheese. In the end, elaborately staged banquets certainly attracted attention, but the occasional promotional dinner and the sale of a few goats for meat ultimately did little to develop an appreciable groundswell of greater public demand for the product.[9]

Not surprisingly, William L. Black had considerable difficulty selling his canned goat meat product. At first he tried to market the cans labeled as "Roast Mutton," but the product did not sell. Black next created a new label for his cans that read "Boiled Mutton," though that didn't work either. Black then had the idea to label his product "W. G. Tobin's Chili con Carne." It's not clear whether Black had permission to use the brand name or simply traded on the past popularity of W. G. Tobin's Chili con Carne, a canned meat product that had been popular years earlier (though Tobin's product did not contain goat meat). Either way, with its new name, Black finally managed to sell his backstock of canned goat meat. After the experience, Black abandoned the meat canning business altogether and passed off the entire ranching enterprise to his sons. This was not uncharacteristic for Black, who throughout his life immersed himself in a variety of enterprises before quickly moving on to the next project. According to Black's daughter Edith, thousands of leftover labels remained at the ranch after her father's death, a reminder of Black's ambitious plans for selling canned goat meat, as well as a testament to the hurdles to be overcome along the way.[10]

William L. Black went on to publish the definitive guide to the Angora goat business of the period, *A New Industry: Or Raising the Angora Goat and Mohair, for Profit* in 1900. The book served as a historical account of the Angora goat in the United States, a how to guide for raising the

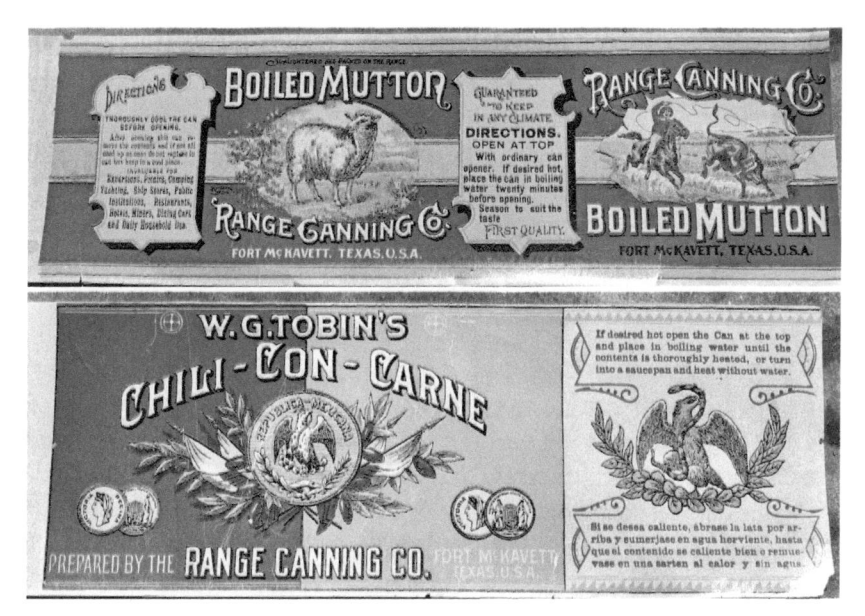

Two of the labels Texas rancher William L. Black used in an attempt to sell his canned goat meat. Collection of author.

animals, and an advertisement for the breed and the industry. Despite his experiences, Black remained, at least on paper, a committed and enthusiastic promoter of goat meat. He insisted that consumer tastes would eventually shift toward goat meat consumption, as the meat was superior to that of sheep. "Angora is now recognized ... as a standard class of meat," said Black, maintaining that the product should be called "'*Angora Venison*' instead of being palmed off for '*Choice Lamb*'" (emphasis in the original). Black believed the public's prejudice against goat meat was a result of the inferior qualities of the common goat. Angora goat meat, he said, "is as superior to the common goat as the Belgian Hare is to our wild native jack rabbit." Despite Black's professed optimism, the term "Angora venison" never caught on with the American public.[11]

❖

After the Texas Sheep and Goat Raisers Association adopted the term *chevon* for goat meat, reaction was mixed. Goat meat boosters hoped that the term would make goat meat more palatable in the minds of the American public. Their renaming efforts resemble the actions of contemporary dairy goat breeders, who imported European dairy breeds from Switzerland, Germany, and France to replace the common, or "scrub," goat. Some felt the French sounding word had potential: "As chevon, [goat meat] should make the social climb into the aristocracy of all things edible," wrote one newspaper editorial. Others were more skeptical. Another newspaper noted wryly, "Chevon is the name given to goat meat by goat raisers. Changing the name may not change the flavor." Either way, the US Department of Agriculture officially accepted chevon as a legally permissible term for goat meat in 1924. But despite the Angora ranchers' initial enthusiasm about the term, in the long run their rebranding efforts backfired. In 1935, the Texas state legislature passed a resolution asking the US secretary of agriculture to officially allow the use of the term mutton in reference to goat meat, claiming that the designation "chevon" had created "a hardship on the sale and marketing [of goat meat]." While no details were provided, it's likely that consumers either had no idea what the name chevon meant or the new name did not affect their feelings about goat meat. The USDA rules remained unchanged.[12]

As the goat meat/mutton debate generated publicity across the country, other groups of goat farmers began to enter the conversation. Amidst the milk contamination scandals of the late nineteenth and early twentieth century (see chapter 3), a goat dairy industry had emerged. Once cow's milk had been shown to be capable of transmitting tuberculosis to humans, an increasing number of consumers turned to goat's milk as a perceived healthier alternative. Dairy goat keeping became a popular and profitable vocation. *The Goat World*, one of the popular dairy goat periodicals of the day, devoted some attention to the meat issue, publishing stories that instructed readers how to cure goat meat and turn it into sausage, as well as offering recipes for cooking goat meat. The publication did not offer any grand ideas about how to persuade the general public to eat goat meat, however. Daniel F. Tompkins, an early dairy goat advocate and the first president of the American Milk Goat Record Association, experimented for a time with several meat preparations including what he called a candy meat made with sugar syrup combined with goat meat, as well as an "Emergency Goat and Kid Meat Biscuit" consisting of goat meat, flour, and butter, which he said kept for several years, though neither product was ever brought to market. For the most part, dairy goat owners focused on developing the market for goat's milk rather than selling their animals for meat. When the dairy goat industry finally hit its stride in the 1980s, that perspective would change.[13]

Meanwhile, communities of immigrants across the country regularly kept goats and consumed their milk and meat during the late nineteenth and early twentieth century. While immigrant families kept goats individually in tenement houses in larger cities, as well as in backyards of small communities, others made their living from livestock ranching. Italian immigrant Giovanni Ammirati owned an Angora ranch in Northern California near Redding during the early twentieth century. Ammirati sold mohair, as well as milk and cheese, and found a ready customer base for goat meat in San Francisco's large Italian population, just a few hundred miles south of his ranch. A short article about Ammirati in San Francisco's *L'Italia* newspaper in 1918 noted, "Every year in January and February, a goat gives birth to two or three kids. The breeder sells the male kids because they are superfluous . . . Giovanni Ammirati [recently] sent 100 kids to Italian and Spanish Butchers in San Francisco, where they are in high demand because of the traditional Easter Celebrations." Likewise, Greek and Italian immigrant entrepreneurs in labor and mining camps in Montana, Colorado, and Utah developed their own goat cheese and meat businesses serving their respective communities. On the East Coast, a similar practice existed among Italian communities in cities including Philadelphia, New York City, and Baltimore. Ads for "Capretti per la Pasqua" (goat kids for Easter) were a regular feature in Italian newspapers. Even as goat ranchers and consumers argued over goat meat marketing tactics, there was already a thriving market for goat meat within America's immigrant communities.[14]

❖

The goat meat/mutton conundrum idled until 1967, when the US Congress passed the Wholesome Meat Act, which was subsequently signed by President Lyndon Johnson. The act was the first significant legislation covering issues related to meat inspection since the Food Safety Act in 1906 and represented a significant advance in meat safety practices. Until its passage, federal laws only regulated meat shipped across state lines; wide disparities in meat inspection standards across individual states had led to significant corruption and abuses, not to mention disparities in labeling requirements. One of the most important aspects of the Wholesome Meat Act, sometimes referred to as the "Equal-To" Act, mandated that state inspection programs be equal to the federal program. The Wholesome Meat Act effectively standardized meat inspection practices throughout the United States.

More importantly for goat raisers, a key section of the act expressly forbade the false and misleading labeling of meat products. The provision reopened old wounds for a new generation of goat ranchers. According to one report, as soon as the act was signed into law, sales and slaughter of goats in Texas virtually ceased, and stockpiles of goat meat began to accumulate in the state's cold storage facilities. The Texas Sheep and Goat Raisers Association sprang into action once again, petitioning the USDA to approve the use of two specific terms, mutton and chevon, for goat meat. The organization reasoned that the word mutton had been used to refer to goat meat for many years—essentially admitting, and asking permission to continue, a practice that public perceived as deceptive. The Texas Sheep and Goat Raisers also argued that it was discriminatory to force them to use the term "goat meat" for its products "because meat from other species of animals does not have to be identified by names that include reference to the common names of the species of animal from which it derived; for example: cattle meat, pig meat, and sheep meat."[15]

While goat ranchers grappled with the future of their industry, their petition generated yet more negative publicity. In a 1971 article "Label Issue Stirs the Goat Raisers," *New York Times* writer William M. Blair posed the following hypothetical: "Will an American eat a hot dog if he knows it contains goat meat?" As the article made the rounds of syndication, the headlines supplied independently by local newspapers became increasingly sensational: the headline in the *Daily Oklahoman* read "Storm Started by Goat Meat in Hot Dogs," while the *Morning Call* in Allentown, Pennsylvania, ran the headline "Hot Dogs? How About Goat Dogs?" Consumers were left to ponder an unsettling image over their morning coffee: the all-American hot dog, contaminated by something they were sure they disliked (goat meat) by those insisting that the perceived deception was perfectly legitimate (goat ranchers). This was not a battle goat ranchers were going to win. The Department of Agriculture eventually announced its decision: the only acceptable terms for labeling goat meat would be "goat meat" or "chevon." This remains the law regarding the labeling of goat meat in the United States.[16]

For many decades, a substantial amount of goat meat sold in the United States was labeled (or mislabeled depending on your perspective) as mutton. The public relations nightmare generated by the perceived deception fostered a deeply negative reputation in the minds of consumers surrounding goat meat. The resulting public conversation centered on the practices of ranchers and marketers rather than any inherent qualities of the animals or their meat. The decades spent debating nomenclature ultimately represent decades lost convincing consumers of the suitability and value of goat meat as a healthy, nutritious product in its own right. Yet despite early industry missteps, demand for goat meat would eventually surge in the United States.

Goat Meat in Contemporary Society

Much has changed since the goat meat labeling debates of years past. At the onset of the Great Depression mohair prices plummeted, forcing many Angora goat ranchers out of business. Government subsidies initiated to support the mohair industry after World War II were eliminated in 1993, sending Angora goat ranching into an even steeper decline. According to the USDA, as of January 2025 there were just 102,000 Angora goats in the United States, substantially less that the total population of over 3.5 million in 1930.[17]

A dedicated meat goat industry has since replaced Angora ranching as the nation's primary source of goat meat. As of 2025, 1.975 million goats, almost 80 percent of the total number of goats in the United States, are being raised specifically for meat. A new generation of breeds have replaced the Angora, many of which have been selectively bred to maximize meat production. One of the more common and recognizable of these is the Boer goat, a breed originally developed in South Africa as a mix of indigenous African goats crossed with European breeds and first introduced into North America in 1993. Boer goats stand out from other goats with their distinct muscular appearance; the breed was specifically developed to gain more muscle than the average goat, and full-grown males can weigh as much as three hundred pounds. There are a number of other goat breeds commonly raised for meat in the United States, among them the Savanna, also from South Africa, and the Kiko, a breed originally developed in New Zealand.

The goat dairy industry has become a significant source of goat meat in the United States. When the goat dairy industry began to expand significantly starting in the 1980s, the population of dairy goats grew along with it. Every year thousands of baby goats, or "kids," are born on goat dairy farms across the country. While some males are kept for breeding purposes or occasionally kept or sold as pets, many if not most are sold for meat. Goat dairy farmers often sell their male offspring to brokers who resell them to individuals or at livestock auctions. In some cases goat dairies or cheesemakers sell animals to individual consumers on a more informal basis. Larry and Clara Hedrich, founders of LaClare Creamery, a goat cheese company in Malone, Wisconsin, started a complementary business called Calanna Specialty Meats in 2018. Calanna raises and sells

Boer goats. Photo by user Lazarus000, Wikimedia Commons, reprinted under Creative Commons license CC BY-SA 4.0.

goats for meat, many of them male goats from the two large goat farms that provide most of LaClare's milk supply.

Although goat ranchers of decades past feared that the transparency mandated by the Wholesome Meat Act would effectively end goat meat sales in the United States, in fact the opposite has occurred. Demand for goat meat has actually risen substantially as waves of immigration brought African and Southeast Asian populations into the United States with immigration policy reform efforts in the 1980s and '90s. Increasingly, a large and growing segment of Americans find goat meat as familiar and desirable as beef or chicken may be to others. The result is that domestic demand has far outpaced the supply of goat meat produced in this country. The population of goats raised for meat in the United States is not even close to meeting current domestic demand.

Why don't United States farmers simply raise more meat goats? Consumers often ask the same question of contemporary clothing retailers or high-tech companies that outsource their manufacturing or production overseas. The answer is the same across industries: it is simply far cheaper to import the product in question, in this case goat meat, from abroad than it is to produce the same product domestically. Currently about 65 percent of goat meat sold in the United States is imported from Australia. That country shipped nearly 14,000 metric tons (over 30 million pounds) of goat meat to the United States, its largest export customer, in 2022.[18]

Australian goat meat is so inexpensive in part because most of the country's goat meat comes from what are called rangeland goats, a long-established wild population descended from animals first introduced by English colonizers in the late eighteenth century. Though the number of rangeland goats is impossible to capture with absolute precision, the total population is thought to be anywhere between 2 and 2.5 million animals at any given time. In the context of the rangeland goat population, ranching is a process of rounding up the feral goats, sometimes termed "opportunistic harvesting." In recent years management programs have been developed for rangeland goats, and some farmer-entrepreneurs fence, feed, and maintain rangeland goats for varying periods of time in the same way a goat farmer might manage a typical domesticated herd.

Sixty-five percent of all the goat meat consumed in the United States is imported from Australia. Author photo.

You might see a business opportunity in raising and selling goats for meat in the United States. Though the idea may be tempting, there are many practical challenges in the business of raising and selling meat goats. One of the most significant is that livestock management practices typically used to manage cattle do not necessarily translate to managing goats. For example, it is difficult to raise large numbers of goats in a large-scale, confined feedlot environment, the way that cattle are typically raised to maximize profit in the industrial beef industry. Goats are active, curious animals and do not thrive in confined fenced environments. Goats in such environments often become stressed and aggressive and become sick and malnourished. Goats are also particularly susceptible to parasites that spread quickly where large numbers of goats are closely confined. These are among the reasons why meat goat farmers keep fairly small numbers of goats. A study conducted in 2019 by the National Animal Health Monitoring System determined that the average goat herd size in the United States (the average included both dairy and meat operations) was just *twenty* goats. By contrast, many industrial cattle operations maintain thousands, even tens of thousands, of animals at a time. While fewer livestock feedlots overall are generally a good thing for the climate and for animal health and welfare, in purely economic terms these issues make goat ranching a more expensive proposition.[19]

Another important issue facing both the goat and sheep ranching industries is the relative lack of available tools to manage animal health. To date there are few medications approved by the US Food and Drug Administration available to farmers to treat small ruminant diseases. The reasons for this are mostly a matter of scale. As of January 2025, there were just under eighty-seven million cattle (both dairy and beef) in the United States, compared to a total population of 2.5 million goats and just over 5 million sheep. The total economic return from cattle products, including both milk and meat, amounted to over $170 billion in 2022. Many, many millions of dollars have been poured into research and development toward maintaining and improving cattle health by industries with a substantial investment in cattle and cattle products. To date, far less investment has been made toward the health and welfare of goats, which are considered (along with sheep) a "minor" species by the FDA due to their comparative lack of economic significance. The Minor Use and Minor Species Animal Health Act, commonly known as the MUMS Act, passed in 2004, attempted to remedy some of these issues by providing incentives for manufacturers and modifying some threshold requirements for use and approval of new drugs for small ruminants. Despite these efforts, not much has changed. From a farming perspective this means goat keeping and care requires significant individual skill, and farmers may face health management challenges that are difficult if not impossible to address adequately.[20]

The potential challenges for meat goat farmers extend past the farm. For over a century the meat-processing infrastructure in the United States developed to facilitate cattle processing. The Covid-19 pandemic exposed flaws in the system that have accumulated over many decades, including industry consolidation; currently the meat-processing industry in the United States is dominated by four large companies. For livestock farmers outside of the mainstream beef and pork industries, processing options are comparatively limited. The relative lack of meat-processing plants means that goat farmers may have to travel long distances to find a suitable facility, which cuts into their bottom line. In addition, the majority of slaughter facilities in the United States are mechanically outfitted to handle larger animals like cows. From the perspective of the average butcher, a 150–200-pound meat goat, or even ten of them, are not worth the time and effort to process (the average cow weighs at least 1,000 pounds or more).

Goat farmer Leslie Svacina has taken a proactive approach to the challenges presented by the goat meat industry. Svacina owns a 140-acre farm in northwest Wisconsin called Cylon Rolling Acres where she raises meat goats. Like farm entrepreneurs of all kinds, Svacina spent her early years in the business selling meat products at local farmers markets and holiday markets. More recently Svacina has turned her focus to direct marketing via the internet. "There are so many costs that go into raising an animal, from feeding to processing," said Svacina. "With direct marketing, the farmer recoups more costs for the product." Through the farm's website, customers can purchase a variety of cuts of goat meat according to preferences or taste; the site also highlights the nutritional characteristics of goat meat and provides a host of recipes for the goat curious. Svacina highlights the extensive work she and her family put into regenerative farming practices, managing goat grazing for pasture health, and working with the farm's ecosystem for the benefit of the animals as well as the environment—values important to today's consumers. Svacina is optimistic about the general market for goat meat. "Consumers are definitely interested in goat meat," she said, "it's an opportunity to connect with a farm and a farmer, rather than the impersonal meat marketplace." Svacina is among an emerging generation of domestic goat farmers who are moving goat farming and marketing into the future.[21]

❖

In the United States, the primary source of consumer demand for goat meat is primarily among Hispanic, African, and Asian communities. Within these communities demand tends to be seasonal; demand rises considerably around holidays such as Eid al-Adha, Christmas, and Easter. In addition, cultural preferences vary as to the age, weight, and sex of the goat to be consumed. Some consumers prefer animals quite young; one common Mexican meat specialty is cabrito, or roasted goat kid; the preferred animal is typically four to eight weeks old. A study of Minnesota's Somali community found that that they prefer meat from an animal of around twenty to forty pounds. For goat farmers, navigating varying demand cycles and other specific requirements can be a complicated prospect.[22]

In addition, some American goat meat consumers require specific meat-handling practices. Many Muslims require their food to be certified as halal, or prepared according to conditions required by Islamic law. Halal requirements include specific slaughtering practices, as well as cleaning and maintenance of equipment. In order to be certified, halal facilities must be inspected and monitored by one of a number of agencies operating in the United States. Comparatively few meat-processing facilities are certified halal or are willing to work within halal requirements. As a result, a goat farmer wanting to direct their meat toward Muslim consumers, may have difficulty locating a suitable processing facility.

Because of the increasing demand for halal goat meat and other halal products in the United States, a halal foods infrastructure has developed to meet the demonstrated and growing demand. One such operation is Illinois-based Halal Farms USA, a slaughterhouse and meat-processing operation which purchases meat goats (among other livestock) from producers that are able to meet its requirements. Halal Farms USA markets its halal products through wholesalers as well as via Billydoe Meats, an online meat retailer. Ayman Noreldin of Billydoe Meats said that Pakistanis and Indians make up the bulk of his customers. For these customers, he says, lamb and goat meat are more an everyday part of their diet. "We also have a lot of Greek and Mediterranean customers," he said, "their buying habits tend to be more holiday centered and seasonal." Even so, Noreldin says that holidays remain the busiest time of year. "We want to promote goat farming," says Noreldin, "there is plenty of demand for goat meat and it can be profitable for us as well as goat farmers."

Over the past several decades the state of Minnesota has seen a sharp increase in the number of Somali immigrants to the state; currently more Somali immigrants live in Minnesota than any other state in the nation. Because the majority of Somalis are Muslim, the state has seen a subsequent rise in demand for halal food of all types, including goat meat. The University of Minnesota has estimated the market for halal goat meat just within that state at $20 million dollars annually. Rising demand is driving the construction of a certified halal meat-processing plant in Willmar, Minnesota (about ninety miles west of Minneapolis), assisted by funding provided by the USDA. The plant, which will be known as Happy Halal, is designed as a halal-specific meat-processing facility that will handle only goats and lambs. Greg Wierschke said dozens of regional goat producers are ready to supply the facility. "We think we can sell upwards of 260,000 goats a year just in Minnesota," says Wierschke. He is cautiously optimistic about the potential for growth in the market for goat meat across the state. "Once people taste good, farm raised goat meat, then I think we can market on that taste." Construction of the facility was set to begin in late 2024.[23]

❖

If you live in the United States and you have tried to shop for goat meat, you know it can be a challenge. Goat meat is not generally available in mainstream grocery stores or butcher shops, though you may occasionally find it in the freezer aisle. At the same time, goat meat is more readily available at shops and restaurants that cater to customers where goat is a familiar part of the culture and cuisine. Larger cities in particular usually have dozens of Mexican, Asian, or halal markets, which likely carry fresh or frozen goat meat. The relative absence of goat meat at the mainstream retail level is reflected in the lack of goat meat dishes offered by mainstream restaurants. The average American restaurant is far more likely to have a goat logo than a goat meat dish on its menu. Despite this, meat in the form of Hmong goat stew, Pakistani nihari, and dozens of other tasty goat meat-centric dishes are readily available at restaurants across the country, if you are willing to seek them out.

The Southwestern United States has long been known for its goat population. Spanish colonialism first brought goats to the region starting in the sixteenth century, and Spanish culture and livestock, including goats, have influenced Western culture and cuisine. One of the higher profile of traditionally goat meat dishes that has come to the fore of public consciousness in recent years is birria, a meaty dish that traces its origins to the Mexican state of Jalisco. Birria is concoction of chiles and spices that are rubbed or stewed, traditionally using goat meat. In Los Angeles, goat birria has become ubiquitous at restaurants and food trucks across the sprawling landscape. Among the higher profile of the region's *birreros* (chefs who specialize in birria) is Juan Garcia, owner of the Goat Mafia food truck based in Compton, California. Garcia is a fourth-generation *birrero* keeping his family's food traditions alive. Born in Jalisco and raised in Compton, Garcia worked a number of jobs before going to culinary school and entering the food business full time. As he likes to put it: "*Si no es chivo, no es birria*"—if there's no goat, it's not birria. He works with a goat rancher north of Los Angeles to get his meat, then prepares the meat with the proprietary spice blend he developed. Birria, he says, is about preserving and maintaining his family's traditions. Garcia says he sees a wide variety of customers at his food truck, from Mexicans hungry for a familiar food to Koreans who tell him that his birria evokes memories of goat dishes from their home country. He told me he doesn't often hear skepticism about goat meat, but if he does, it's from those who are unfamiliar with it.

Juan Garcia barbecuing cabrito (young goat) at his Goat Mafia food cart in Los Angeles, California. Author photo.

Goat meat is also commonly found in many Caribbean cuisines. Columbus first brought livestock to the Caribbean on his second voyage to the region, which he called the West Indies, in 1493. Over the next several decades the Spanish spread their influence across the Caribbean islands, including Jamaica. The Spanish were particularly interested in Jamaica because of its proximity to South America, their next colonization target. Curry spices made their way to the Caribbean after England abolished slavery in the eighteenth century and Indian indentured servants migrated to the Caribbean to work in the sugar industry. Over the centuries curry became integrated into the cuisines of the Caribbean.

Caribbean-born immigrants make up more than 10 percent of the United States foreign-born population, and as Caribbean communities across the United States have grown, Caribbean cuisines have spread as well. One goat dish found among many of the Caribbean islands is curry goat, a flavorful slow-cooked goat stew. In Jamaica, curry goat is often a part of Sunday dinners or celebrations like weddings, graduations, and festivals. Yanikie Tucker, author of the book *70 Jamaican Recipes: From My Grandmother's Kitchen*, says there are two keys to making a great Jamaican curry goat, "it must have Scotch Bonnet peppers and Pimento berries (also known as allspice)." Jamaican curry goat is rich, flavorful, and a bit spicy, but for those interested in exploring new and unfamiliar goat dishes, curry goat can be a great introduction to the wide world of goat meat.[24]

The Argument for Goat Meat

Though there is a substantial and increasing demand for goat meat in the United States, a large segment of the American public does not, or will not, eat goat meat. Some are repulsed by even the thought of consuming it. Many who have tasted goat meat consider it tough or gamey. Some of the more negative reactions make a certain sense, since few Americans of European heritage were raised eating goat meat and their food or cultural traditions largely do not include goat meat. And for those unfamiliar with goat meat there is no immediate reason to like goat meat, or consider eating it, when there is plenty of readily available beef, chicken, and pork available across the United States. Consumers who are unaccustomed to goat meat and rarely, if ever, see it in a grocery store or on a restaurant menu are essentially stuck in an infinite feedback loop where goat meat may as well not exist at all.

Despite the skepticism about goat meat that persists among part of the American population, studies have consistently shown that a distaste for goat meat does not necessarily bear out in a controlled setting. One 2005 study showed that consumers across ethnicities were generally willing to try goat meat. Another study showed that, among survey respondents, only 2.9 percent said they would never consume goat meat under any circumstances. One study that consisted of blind tastings of beef, pork, and goat meats, showed that participants viewed goat meat favorably as compared to the other samples. "However novel the idea of eating goat might be," researchers concluded, "the actual eating experience is rather ordinary."[25]

During the late 2000s, goat meat seemed poised to become the next big food trend. The buzz was generated in part by news of Bill Niman's turn to goat ranching in 2008. Niman is a former back-to-the-lander who started raising a few cows and pigs in the 1970s on a small Northern California ranch, selling his pork to Alice Waters's restaurant, Chez Panisse. Niman Ranch grew into a high-profile cattle ranching business over the years, and after a series of expansions and buyouts, Niman left his namesake ranch in 2007. When Niman took up goat ranching, many thought his deep industry knowledge and considerable credibility in meat-business circles could lead goat meat into the mainstream of meat consumption in the United States. Niman's goat meat promised to be a boutique organic, local product, exactly the type of meat appreciated by chefs, restaurateurs, and consumers seeking food products in those categories. Just as goat's milk cheese was catapulted into the mainstream as a gourmet product during the 1980s, some felt that goat meat was destined to follow—and this appeared to be its time to shine. Reporter and food writer Kim Severson was optimistic about goat meat's potential: "If you put [goat meat] in the hands of high end chefs, you end up with a nice dish, and it trickles down. A lot of food trends started this way." Within a few years, however, Niman sold his goat ranching business and the buzz faded. Since then, periodic waves of media coverage repeatedly predict that goat meat is about to go mainstream in the United States. Inevitably, star chefs are quoted, the health virtues of goat meat are expounded upon, and delicious recipes are hyped, but to date the promised goat meat revolution—the introduction of goat meat into mainstream American consciousness and cuisine—has not materialized.[26]

Whatever the perceived negative qualities of goat meat, it actually possesses a number of virtues typically valued by consumers. For shoppers who are paying close attention to fat and protein content of food products, goat meat stands out. Goat meat is leaner than beef and lower in cholesterol and saturated fat than other commonly consumed meat products. In addition, goat meat contains fewer calories per serving than beef. Goat meat produced in the United States is also likely to have been raised on pasture, a feature considered a high virtue in beef or pork products. For those concerned about the ethics and sustainability of their food choices, goat meat can be a good option.[27]

These days contemporary consumers are focusing on the climate and carbon impact of all aspects of daily life, including fuel and electricity consumption as well as the carbon footprint of agricultural products. One of the central points of concern in the climate change conversation is methane production in the livestock industry. Both humans and animals produce methane as a byproduct of the digestive process, and goats are no exception. In fact, studies have shown that goats produce *more* methane per pound than cattle. At the same time, studies have also shown that farmers can manage the problem by feeding goats specific diets developed so that they will produce less methane. In addition, where goat manure is managed and applied to pastures (as cow manure is in mainstream dairy operations), the overall effect is net carbon sequestration. Others argue that corporate interests encourage discussion about livestock methane production to mask far more serious problems posed by widespread industrial pollution.[28]

Goats may become advantageous in a climate changed world. Goats are genetically predisposed to thrive in the hot, dry conditions forecasted to become increasingly common. One promising area of research in the area of climate-friendly livestock management practices has centered around the idea of co-grazing, a system in which a variety of species graze alongside one another in the same pasture environment, reflective of the way pastoral cultures across the world have been grazing livestock alongside one another for centuries. Multispecies grazing creates efficiencies because different species prefer different plants and thus do not compete for food. Goats, for example, prefer to graze on leaves and brush rather than grass (though they will eat that as well), while cattle and sheep consume only grass. The process of combining grazing species has been shown to increase biodiversity and carbon sequestration. From a farming perspective, co-grazing is a win because the economic productivity gained on one segment of land increases by grazing more animals of different species together, rather than fewer animals of just one species.

Any discussion about modern-day meat consumption must also contend with the new world of meat alternatives. Just as sales of plant-based dairy products have soared in the last decade, so too have sales of plant-based "meat" products. United States producer Beyond Meat reported a net revenue of $343.4 million for 2023; its main United States competitor, Impossible Foods (not publicly traded) is estimated to have generated $256 million in revenue in the first half of 2023. Though plant-based meat sales have slowed considerably since their initial entry into the marketplace, a newer generation of startups are developing a meat alternative produced entirely from real animal cells grown in a controlled laboratory environment. So-called cultured meat products take the discussion of meat alternatives into another realm entirely. One of the main arguments made in favor of cultured meat is that the products possess a lower carbon footprint compared to real-world livestock raising and meat production. Early studies showed that cultured meat possesses a smaller carbon footprint than conventionally produced meat, supporting the industry's claims of green virtuousness and validating the infusion of millions of dollars into continuing research and development. That being said, in 2023, a preliminary study (not yet peer reviewed) by researchers at the University of California, Davis showed that the growth media used to culture animal cells carries a carbon footprint as much as 25 percent *higher* than that of farmed beef. The science is complicated and continues to evolve.

For now, it remains to be seen to what extent consumers will accept or consume cultured meat products, or whether any company will bring a cultured goat meat product to market in the United States. That being said, cultured meat products are legal and are currently being sold (and consumed) at restaurants in Singapore. In the meantime, the states of Florida and Alabama have banned the manufacture and sale of cultured meat within their borders. Increasingly, however, what is clear is that terms of the global meat consumption conversation are rapidly evolving. Goat farmers, as well as livestock ranchers of all types, would do well to take notice.[29]

❖

In almost any industry, food-based or otherwise, when a product is introduced to the marketplace, a marketing plan accompanies the launch. Perhaps a trade group might develop strategies to sell the product and lobby politicians both locally and nationally to help further its members' interests and bottom line. Although there are a number of meat goat-specific breed registries in the United States, including the American Boer Goat Association, the American Kiko Goat Association, and others, to date there is no industry-specific trade organization devoted specifically to promoting and marketing domestically produced goat meat as there are organizations devoted to beef, pork, and lamb. Like the American goat dairy industry, the goat meat industry lacks a marketing infrastructure. Agricultural universities and their associated extension programs are, generally speaking, the only entities that devote much attention and resources to bridging the considerable gap between goat producers and the marketplace, but their efforts are primarily directed at educating and assisting farmers rather than toward developing the broader industry initiatives.

In the United States, agricultural commodities of all types have long benefited from the assistance of entities called marketing boards, industry-specific organizations that operate under the auspices of the USDA's Agricultural Marketing Service. Congress allocates funds to support the efforts of marketing boards, and the boards develop programs to promote their product to consumers. Some states conduct regional programs of their own. Marketing board programs can take many forms, such as restaurant specific initiatives or nationwide consumer awareness campaigns. Familiar slogans like "Beef: it's What's for Dinner" and "Pork: the Other White Meat," highly successful marketing slogans of decades past, are representative of marketing-board promotional efforts. In 2023, the Lamb Marketing Board developed a number of initiatives, including a grilling promotion in conjunction with a major national grocery retailer promoting the consumption of American lamb burgers. If the domestic goat meat industry were able to organize and access this funding, the positive promotion generated by such a comprehensive marketing campaign could go a long way toward raising consumer awareness about the virtues of goat meat, particularly domestically produced goat meat.

Despite the general dearth of marketing and promotion efforts, there have been grassroots campaigns devoted to promoting goat meat consumption. One of the more well-known of these is Goattober, a month-long promotional event started in 2011 by New York cheesemonger Anne Saxelby as a way of addressing the problem of male goats in the goat dairy industry in the New England area. The event, still held annually in October, is designed to highlight goat meat and encourage its consumption by enlisting restaurants to feature and highlight dishes made with goat meat. In its early years, the Goattober concept spread to countries across the world, including Japan, Australia, and the United Kingdom. Although the celebration lost considerable momentum during the Covid-19 pandemic as numerous restaurants closed their doors, some United States goat farmers and restaurants continue to spread the word about goat meat and other goat products during the month of October.

❖

Goat meat has a long and fraught history in the United States. Goats have been consumed for meat since European colonists first brought the animals to the Western Hemisphere centuries ago, though they were generally cast aside in favor of beef cattle. Waves of immigrants to the United States in the nineteenth and early twentieth century continued the practice of keeping and consuming goats for meat as well as milk, though the practice was frowned upon by many more established many others. Even when Americans began to embrace goats for their milk production during the early twentieth century, that appreciation did not generally extend to goat meat. And the marketing practices of Angora goat ranchers during the early twentieth century, which the general public perceived as deceptive, only deepened the already negative cultural assumptions about goat meat.

While discussions about goat meat have generally moved past deceptive labeling issues, goat meat continues to occupy a relatively unique position in the broader American food landscape. While goat meat has failed to gain traction with some American consumers, it is actually more popular now than it has ever been in American history—one person's "gamey" is another person's "flavor." And over a century after the term chevon was first devised as a dressed up, French sounding word for goat meat, it must be said that few outside of the ranching and food and beverage industries are familiar with the term. In 2009, a *New York Times* writer said that word chevon sounds like "a miniature Chevrolet or a member of a 1960s girl group." It's time to move on from chevon.[30]

But there may yet be hope. American consumers are becoming more adventurous in their eating habits, and the increasing availability of a wide range of world cuisines in restaurants, food carts, grocery freezer cases, and online retailers has expanded American minds and palates. No one living in the United States in the 1920s would have imagined that Indian cuisine would be widely available across the United States a few decades into the future or that their descendants would be eating Thai curry with coconut milk on a regular basis. In contemporary society, social media has become its own marketing and promotional medium, as food influencers on every platform promote recipes and provide insight into their country's cuisines. Perhaps a few decades from now, goat meat dishes will be as ubiquitous as beef hamburgers or fried chicken is currently. Goat ranchers certainly hope so.

Acknowledgments

Thank you to Oregon State University Press, and especially to Acquisitions Editor Kim Hogeland for her willingness to take on a book about goats, of all things. So many people provided valuable information, advice, and assistance as I was writing this book, including Jonathan Kauffman, Anna Thompson-Hadjik, Ken McMillin, Ricki Carroll, Janet Fletcher, Miles Hooper, Gianaclis Caldwell, and Milena Anfosso, and so many others. Thanks also to cheesemaking pioneers Jennifer Bice, Mary Keehn, Judy Schad, and Allison Hooper, who generously took the time to speak with me about the early years of American artisan goat cheese production. Special thanks to Jeff Roberts, whose enthusiasm about the history of goats helped me feel like this project was worth taking on in the first place. Also, thank you to Pat Morford of Rivers Edge Chevre. When I first became interested in the cheese world during the early 2000s, Pat's farm along the Oregon coast was the very first I visited. Her warm welcome (and fresh chevre) was my entry into the wide world of cheese. Pat was also an early supporter of my idea of a book about goat history. In addition, this book could not have been written without the librarians, archivists, and special collections department folks who I've encountered along the way; your kindness and generosity make research fun. Thanks also to editor Elena Abbott for her editorial advice and encouragement, not to mention her enthusiasm about the topic of goats. Most of all, many thanks to all of the kind goat farmers, cheesemakers, industry experts, and others who were generous enough to take the time to talk with me.

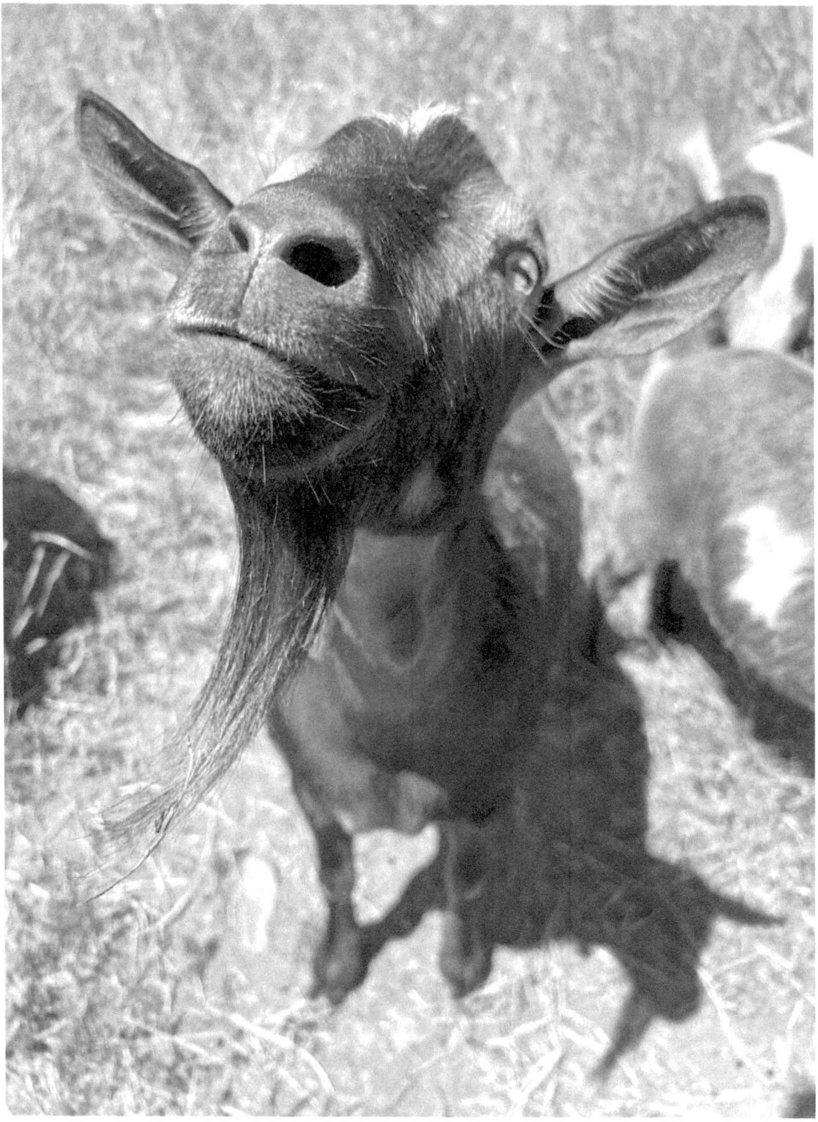

Notes

Chapter 1: Poor Man's Cow

1. The goat's obituary appeared in several London newspapers, including *Boddely's Bath Journal* April 6, 1772, and *The Public Advertiser*, April 3, 1772. See George Birkbek Hill, ed., *Boswell's Life of Johnson* (MacMillan and Co., 1887), 2: 144, for the poem; "A Chance for Lawson N. Fuller," *New York Times* April 6, 1890 (epigraph).

2. Berthold Fernow, ed., *The Records of New Amsterdam 1653–1674 Anno Domini*, Vol. I., *Minutes of the Court of Burgomasters and Schoepens 1653–1655* (Knickerbocker Press, 1897), 405–406; Arnold J. F. Van Laer, trans., *New York Historical Manuscripts: Dutch*, Vol. IV, *Council Minutes 1638–1649* (Genealogical Publishing Co., 1974), 123.

3. Fernow, *The Records of New Amsterdam 1653–1674*, 1:3, 16.

4. Sydney V. James Jr., ed, *Three Visitors to Early Plymouth* (Applewood Books, 1997), 25 (Altham); Everett Emerson, ed., *Letters From New England: The Massachusetts Bay Colony, 1629–1638* (University of Massachusetts Press, 1976) (Higginson); Edward Arber, ed., *Captain John Smith: Works 1608–1631, Part II* (Archibald, Constable and Co., 1895) 609 (list of necessities for travel to Virginia).

5. Ralph Hamor, *A True Discourse of the Present State of Virginia* (John Beale for William Welby, 1615), 23; John Farrer, *A Perfect Description of Virginia* (Peter Force, 1837); J. Franklin Jameson, ed., *Johnson's Wonder Working Providence 1628–1651* (Charles Scribner's Sons, 1910) 211, Virginia DeJohn Anderson, *Creatures of Empire: How Domestic Animals Transformed Early America* (Oxford University Press 2006), 104.

6. John Josselyn, *An Account of Two Voyages to New-England* (William Veazie, 1865), 147; Adrian van der Donck, *Description of the New Netherlands*, trans. Hon. Jeremiah Johnson (Directors of the Old South Work, 1896), 47; see also Gervase Markham, *The English Housewife*, ed. Michael Best (1615; McGill-Queen's University Press, 1994).

7. Roger Clapp, *Memoirs of Roger Clapp: 1630* (David Clapp Jr., 1844), 42; Alexander Young, *Chronicles of the First Planters of the Colony of Massachusetts Bay from 1623 to 1636* (Charles C. Little and James Brown, 1846), 415; Clayton Colman Hall ed., *Narratives of Early Maryland, 1633–1684* (Barnes & Noble, 1910), 60. According to historian Carl Bridenbaugh, colonial Rhode Island farmers' wives made goat's milk cheese: see *Fat Mutton and Liberty of Conscience, Society in Rhode Island, 1636–1690* (Brown University Press 1974), 42.

8. Sidney H. Miner and George D. Stanton Jr., eds., *The Diary of Thomas Minor, Stonington Connecticut, 1653–1684* (Day Publishing, 1899), 25, 33–40, 51; see also Virginia DeJohn Anderson, "Thomas Minor's World: Agrarian Life in Seventeenth Century New England," *Agricultural History* (Fall 2008): 496–519.

9. John J. McCusker and Russell R. Menard, *The Economy of British America, 1607–1789* (University of North Carolina Press, 1985), 103, 136.
10. Percy Wells Tidwell and John Falconer, *A History of Northern Agriculture, 1620–1840* (Peter Smith, 1941) 32. Archaeologist Craig S. Chartier reviewed the probate records of Plymouth Colony, which detail the gradual disappearance of goats from farm stock as estates were passed between families over the early decades of the life of the settlement. See "Plymouth Colony Livestock 1620–1692," http://www.plymoutharch.com/wpcontent/uploads/2010/12/Plymouth_Colony_Livestock.pdf; *Records and Files of Quarterly Courts of Essex County, 1636–1656* (Essex Institute, 1911), 1:18, 69; Lucius R. Page, *History of Cambridge, Massachusetts, 1630–1877, with a Genealogical Register* (H. O. Houghton and Co. 1877), 41; Charles Hoadley, ed., *Records of the Colony and Plantation of New Haven from 1638–1649* (Tiffany and Co., 1857) 187.
11. *Second Report of the Commissioners of the City of Boston: Containing the Boston Records, 1634–1660, and the Book of Possessions* (Rockwell and Churchill, 1881), 68; Librarian of the Rhode Island Historical Society, ed., *The Early Records of the Town of Warwick* (E. A. Johnson and Co., 1926), 52–53 (Warwick); Jameson, *Johnson's Wonder Working Providence*, 73.
12. In Maryland, for example, it became standard policy for a landowner to require a tenant farmer to plant and tend an orchard as a condition of their tenancy; see Lois Green Carr, Russell R. Menard, and Lorena S. Walsh, *Robert Cole's World: Agriculture and Society in Early Maryland* (University of North Carolina Press, 1991), 35–36; see also Sarah Hand Meacham, "Thy Will be Adjudged by Their Drink What Kind of Housewives They Are: Gender, Technology and Household Cidering in the Chesapeake, 1690–1760," *Virginia Magazine of History and Biography* 111, no. 2 (2003): 117–150.
13. Christopher Dyer, "Alternative Agriculture: Goats in Medieval England," in *People, Landscape and Alternative Agriculture: Essays for Joan Thirsk*, ed. R. W. Hoyle, Agricultural History Review Supplement Series, no. 3 (British Agricultural History Society, University of Exeter, 2004), 20. See also R. E. Prothero, "Landmarks in British Farming," *Journal of the Royal Agriculture Society of England*, 3rd ser., vol. 3 (1892): 4–30.
14. *Daily Alta California*, December 14, 1863; *Daily Alta California*, October 13, 1859; *Daily Alta California*, April 11, 1864.
15. "Local Matters," *The Sun* (Baltimore), April 23, 1857; "Savagely Butted by a Billy Goat," *Philadelphia Inquirer*, July 19, 1899; "The Board of Health and The Goats," *Evening Star*, November 16, 1872; "Goats a Nuisance," *The Republic*, January 27, 1851.
16. For more about the intersection of livestock and growing nineteenth century cities see Andrew Robichaud, *Animal City: The Domestication of America* (Harvard University Press, 2019), 15–17; see also Catherine McNeur, *Taming Manhattan: Environmental Battles in the Antebellum City* (Harvard University Press, 2018); Catherine Brinkley and Domenic Vittello, "From Farm to Nuisance: Animal Agriculture and the Rise of Planning Regulation," *Journal of Planning History*, 12, no. 2 (May 2014): 113–115.
17. *Daily Alta California*, February 28, 1850.
18. *Journal of the Sixth Session of the Legislature of the State of California* (California State Government, 1855); *Daily Alta California*, June 5, 1851; *Daily Alta California*, May 8, 1851. On the practice of goats herded with sheep, see for

example, Frederick Law Olmstead, *A Journey Through Texas* (Dix, Edward and Co., 1857), 258; "City Items," *Daily Alta California*, November 21, 1850 (notes sheep arriving in town "with a sprinkling of goats" and currently kept in "The Square," probably Union Square).

19 "Common Council," *Daily Alta California*, May 4, 1855; *Minutes of the Common Council of the City of New York, 1784–1831* (City of New York 1917), 3:464; *Ordinances of the Corporation of the City of Philadelphia* (Moses Thomas, 1812), 121.

20 "A Nuisance," *Baltimore Sun*, May 1, 1846.

21 "Common Council," *Daily Alta California*, May 4, 1855; "Board of Supervisors, May 6," *Daily Alta California*, May 7, 1857; "New York City," *New York Times*, December 4, 1855; "Somewhat Too Much of Goat," *New York Times*, Nov 30, 1884.

22 "Brutal Pound Deputies," *Daily Alta California*, May 27, 1891; "Lindo Out of Office," *San Francisco Chronicle*, February 3, 1887.

23 "Tackled a Goat," *Philadelphia Inquirer*, December 6, 1896; "Flatbush Police Defeated," *New York Times*, December 5, 1897; "Capture of Twenty Eight Goats," *New York Daily Herald*, June 12, 1874.

24 "Telegraph Hill," *Daily Alta California*, October 16, 1857; Charles Warren Stoddard, *In the Footprints of the Padres* (A. M. Robertson, 1912), 45–46.

25 William Mather, *Geology of New York, Part I* (Carroll and Cook, 1843), 597; "That Washington Slab," *The Sun*, September 8, 1898; "Our Squatter Population," *New York Times*, July 15, 1867; "Dutch Hill," *New York Times*, March 21, 1855.

26 "Life in the Cliff Dwellings," *San Francisco Call*, July 19, 1891; "Hoodlum Gangs," *San Francisco Examiner*, May 6, 1888.

The presence and consumption of goats within the Chinese community in San Francisco is difficult to assess. I found a few scraps of evidence that the Chinese may have kept or consumed goats, though due to the intense prejudice of the period it is hard to know how true the accounts may be. See, for example, Frank Soulé, John H. Gihon, and James Nisbet, *The Annals of San Francisco: A Summary of the First Discovery, Settlement, Progress and Present Condition of California* (D. Appleton and Co., 1855), 384 (discussion of periodic celebrations held by Chinese, including processions, fireworks, and roast goat and pig); "How Chinese Prepare Meat for Market," *San Francisco Call*, October 31, 1897; "Goat Meat is Sold as Mutton," *San Francisco Chronicle*, October 18, 1899. Two experts I consulted said it is unlikely that the Chinese community in San Francisco would have consumed much goat meat. One reasoned that most of the Chinese immigrants in California came from Guangdong Province, an area not known for its goat consumption.

27 "Telegraph Hill Neighborhood Association," *San Francisco Call and Post*, August 22, 1904. See also John T. Appel, "From Shanties to Lace Curtains: The Irish Image in *Puck*, 1876–1910," *Comparative Studies in Society and History*, 13, no 4 (October 1971): 365–375.

28 Charles Loring Brace, *The Dangerous Classes of New York and Twenty Years Work Among Them* (Wynkoop & Hallenbeck, 1872), 160–161.

29 "The 'High Rock Gang' No More," *The Sun*, April 17, 1898; *The Times* (Philadelphia), June 21, 1901.

30 For more on the evolution of zoning and its connection to, and effect on, animal nuisances see Brinkley and Vittello, "From Farm to Nuisance: Animal Agriculture and the Rise of Planning Regulation," 113–135.
31 Complaints from Box 77, Folders 5–9, James Rolph Jr. Papers, MS 1818, California Historical Society Collection, Stanford University Library, Palo Alto, CA.
32 Hettie Belle Marcus, "Lombard Steet," in "Julia Morgan: Her Office and a House," ed. Suzanne Riess, ed., Julia Morgan Architectural History Project, vol. 2 (unpublished manuscript, Bancroft Library, University of California, Berkeley), 139a (describes the neighborhood goats); David Myrick, *San Francisco's Telegraph Hill* (Howell North Books, 1972), 120. Accounts of the Telegraph Hill "goat ladies" can also be found in the memoirs of Kenneth Rexroth, who lived in the area during the 1920s; see "Bohemian San Francisco Between the Wars," https://www.foundsf.org/index.php?title=Bohemian_San_Francisco_Between_the_Wars; "Goats Raid House Porches," *San Francisco Chronicle*, November 27, 1922.

In March of 1928, the *San Francisco Examiner* ran a story noting that city officials had "banished" Millanella Cosenza's goats due to neighbors' complaints. It's not clear whether the date of their official exile (April 1) is an indication of the article's veracity, but I was not able to locate records of any complaints lodged against Cosenza in the city's Board of Health records, nor records of any other administrative action taken by the Board of Health against her specifically. There is also no record of any law specifically addressing goats in the city passed after the 1920 goat ordinance. See Ernest Hopkins, "Tragedy Shrouds Telegraph Hill: Historic Goats and Artists Routed," *San Francisco Examiner*, March 20, 1928.

Chapter 2: Navajo Goats

1 The Diné creation story is an oral history tradition transmitted through the sacred Blessingway ceremony. Among the written accounts of Blessingway are Leland C. Wyman, *Blessingway*, (University of Arizona Press, 1970), which details several versions of the creation story; Aretta Begay on *Toasted Sister*, podcast, episode 31, "Navajo Sheep—They're My Life . . . I Love Them," April 17, 2018; "Navajos Use All of Goat But Smell, Senators See," *Gallup Independent and Evening Herald*, May 18, 1931 (epigraph).
2 Elinore M. Barrett, *Conquest and Catastrophe: Changing Rio Grande Pueblo Settlement Patterns in the Sixteenth and Seventeenth Centuries* (University of New Mexico Press, 2002); Peter Iverson, *Diné: A History of the Navajo* (University of New Mexico Press, 2002), chapter 1; Garrick Bailey and Roberta Glenn Bailey, *A History of the Navajo: The Reservation Years* (School of American Research Press, 1986), 12.
3 Katherine A. Spielmann, Tiffany Clark, Diane Hawkey, et. al., " . . . Being Weary, They Had Rebelled": Pueblo Subsistence and Labor Under Spanish Colonialism, *Journal of Anthropological Archaeology*, 28, no. 1 (2009): 102–125; Robert MacCameron, "Environmental Change in Colonial New Mexico," *Environmental History Review* 18, no. 2 (Summer 1994): 17–39. See also Heather Trigg, *From Household to Empire: Society and Economy in Early Colonial New Mexico* (University of Arizona Press, 2005).
4 On the earliest Navajo herding practices, see James Wade Hadley Campbell, "Exploring The Early Navajo Pastoral Landscape: An Archaeological Study of

(Peri)Colonial Navajo Pastoralism from the 18th to the 21st Centuries AD" (PhD diss., Harvard University, 2021), chapter 6. As Campbell notes, the fact that Navajo pastoralism was not imposed from the outside makes Navajo herding practices a unique form of innovation. See also, Iverson, Diné, chapter 1, and Klara Kelly and Harris Francis, *A Diné History of Navajoland* (University of Arizona Press, 2019). Census numbers from Bernardo de Miera y Pacheco map/census of 1758; numbers rounded to convey scale. Because the Spanish typically referred to sheep and goats collectively as *ganado menor* (small livestock), as was the case in this census, Miera's census includes both sheep and goats, though the numbers are most often translated as simply "sheep."

5 Ron Garnanez on *Toasted Sister*, podcast, episode 31, "Navajo Sheep: They're My Life ... I Love Them," published April 17, 2018, https://soundcloud.com/toastedsisterpodcast/e31-navajo-sheep-theyre-my-life-i-love-them.

6 Bailey and Bailey, *History of the Navajo*, 92; *Annual Report of the Commissioner of Indian Affairs* (Government Printing Office, 1914), 17.

7 Denis Foster Johnston, "Trends in Navaho Population and Education," in David F. Aberle, *The Peyote Religion Among the Navajo* (University of Chicago Press, 1982), appendix A; Bailey and Bailey, *A History of the Navajo*, 41, 133. Livestock population estimates from Bailey and Bailey, *A History of the Navajo*, appendix A.

8 Charlotte Frisbie, with Tall Woman and assistance from Augusta Sandoval, *Food Sovereignty the Navajo Way: Cooking With Tall Woman* (University of New Mexico Press, 2018), 14–16. See also Flora T. Bailey, "Navaho Foods and Cooking Methods," *American Anthropologist* (April–June 1940): 270–290, which contains recipes for thin corn griddle cakes, boiled white corn tamales, and green corn, all of which use milk as part of their preparation. See also James Beadele, *Five Years in the Territories* (National Publishing Co., 1873), 547, which describes an encounter with the Navajo tribe and their use of goat's milk.

9 Nanabah Begay, Oral History, Chinle Agency, Envelope 100, Office of Navajo Economic Opportunity Collection, Navajo Nation Library, Window Rock, AZ; Frank Mitchell, Navajo Blessingway Singer: The Autobiography of Frank Mitchell 1881–1967, eds. Charlotte Frisbie and David P McAllester (University of New Mexico Press, 1978), 32; Edward Sapir and Harry Hoijer, eds., Navajo Texts (AMS Press, 1975), 411; The technique of forming dairy curds into balls resembles the process of making Dutch cheese, a common household cheese in the nineteenth-century United States.

10 According to a source cited by G. F. W. Haenlein in "Dairy Goat Industry of the United States," *Journal of Dairy Science* 64 (1981): 1288–1304, just 413 goats were imported from Switzerland directly into the United States after 1900. Period costs gleaned from contemporary goat journals.

11 Letter from Navajo Trading Company, Cartons 1–2, Charles Frederick Fisk Papers, BANC MSS 70/110 c, Bancroft Library, University of California, Berkeley; "Beware of Island Goats," *Angora and Goat Milk Journal*, July 1920, 33; "Navajo Goats As Foundation," *Angora and Goat Milk Journal*, April 1921, 47; "Goat No Longer a Joke," *Boston Globe*, May 15, 1921 (photo of Rosie accompanies the article); classified ads, *Farmington Times-Hustler*, September 16, 1920; "Local Livestock Market," *Kansas City Journal*, November 30, 1899; *Grand Valley Times* (Moab, UT), October 5, 1917.

12 US Census Bureau, Twelfth Census of the United States, vol. 5 (1900), "Agriculture on Indian Reservations," 732, https://www2.census.gov/library/publications/decennial/1900/volume-5/volume-5-p9.pdf.

13 The Zeh report is reprinted in US Senate, Committee on Indian Affairs, *Navajos in Arizona and New Mexico* (1932), 9121–9132 (quotations on p. 9126; "goat haters" quotation from John G. Hunter, Superintendent of the Southern Navajo Jurisdiction, 9121).

14 "Five Westerners Tell of Life in the Forest Service," *The Smoke Signal*, Fall 1967, 140; Joseph Howell Jr., "The Navajo Sheep Herder," in T*he 1933 Ames Forester* (Forestry Club of Iowa State College, 1933), 48–53. The history of western range research and its influence on the development of environmental policy in relation to the Navajo deserves more examination. See, for example, James A. Young, "Range Research in the Far Western United States: The First Generation," *Journal of Range Management* 53, no. 1 (January 2000): 2–11; James A. Young and Charlie D. Clements, "Range Research: The Second Generation," *Journal of Range Management* 54, no. 2 (March 2001): 115–121; Division of Range Research, Forest Service, USDA, "The History of Western Range Research," *Agricultural History Society* 18, no 3 (July 1944): 127–143.

15 US Senate, Committee on Indian Affairs, *Navajos in Arizona and New Mexico*, 9121 (meat), 9685; Wheeler tells another shepherd, who keeps about 140 goats, the same thing at 9556.

16 "Navajos Use All of Goat But Smell, Senators See," *Gallup Independent and Evening Herald*, May 18, 1931; Broderick Johnson and Ruth Roessel, *Navajo Livestock Reduction: A National Disgrace* (Navajo Community College Press, 1974), 39. The comments of Mrs. Kee McCabe used in the header of this chapter were spoken at the demonstration event and reported in the *Gallup Independent*.

17 David Siddle, "Goats, Marginality and the 'Dangerous Other,'" *Environment and History* 15, no 4 (November 2009): 523 (French history); C. Kieko Matteson, "'Bad Citizens' with 'Murderous Teeth': Goats into Frenchmen, 1789–1827," *Journal of the Western Society for French History* 34 (2006): 147–161. Siddle's analysis goes into fascinating detail about the deeper origins of Western goat prejudice, including centuries old superstitions about good and evil, witchcraft, and associations between goats and satanism.

18 Diana Davis, "Potential Forests: Degradation Narratives, Science and Environmental Policy in Protectorate Morocco, 1912–1956," *Environmental History* 10, no 2 (April 2005): 211–238; Sir A. Daniel Hall, *The Improvement of Native Agriculture in Relation to Population and Public Health* (Oxford University Press 1936), 53; "Erosion in Wakamba Land," *Indians at Work*, March 1, 1934, 13.

19 "Start Thinning Navajo Flocks," *Santa Fe New Mexican*, July 18, 1934; "Erosion and Range Control and Stock Reduction," in *Navajo Tribal Council Resolutions 1920–1951* (Government Printing Office, 1952), 258–260.

20 Johnson and Roessel, *Navajo Livestock Reduction*, 110, 94, 133; Iverson, Diné, 153.

21 Johnson and Roessel, *Navajo Livestock Reduction*, 155, 94, 141.

22 Population numbers from Bailey and Bailey, *History of the Navajos*, appendix A; *Congressional Record, 81st Congress, 1st Session, Vol 95, Part 8* (Government Printing Office, 1949), 10643; Jim Counselor, letter to the editor, *Farmington Times Hustler*, April 3, 1936.

23 For a far more comprehensive treatment of the complex history of the livestock reduction period see Marsha Weisiger, *Dreaming of Sheep in Navajo Country* (University of Washington Press, 2011).
24 Peter Iverson, ed., *For our Navajo People, Diné Letters, Speeches, and Petitions, 1900–1960* (University of New Mexico Press, 2002), 238; "Sheep, Land and People: What Hope for the Future?," *Navajo Times*, April 6, 1978.
25 *Land Reform in the Navajo Nation: Possibilities of Renewal For Our People* (Diné Policy Institute, Diné College, 2017), discussion of grazing practices beginning on 37.
26 Peter McDonald, "Chairman's Spotlight on Food," *Navajo Times*, April 16, 1981.
27 Devon A. Mihesuah and Elizabeth Hoover, ed., *Indigenous Food Sovereignty in the United States: Restoring Cultural Knowledge, Protecting Environments and Regaining Health* (University of Oklahoma Press, 2019), xiii; Marla Pardilla, Divya Prasad, Sonali Suratkar, Joel Gittelsohn, "High Levels of Household Food Insecurity on the Navajo Nation," *Public Health Nutrition* 17, no. 1 (2014):58–65.
28 *Diné Food Sovereignty: A Report on the Navajo Nation Food System and the Case to Rebuild a Self-Sufficient Food System for the Diné People* (Diné Policy Institute, 2014); Denisa Livingston, speaking at Sheep is Life Conference, Window Rock, Arizona, June 16, 2023.
29 A. Park Williams, Benjamin I. Cook, and Jason E. Smerdon; "Rapid intensification of the Emerging Southwestern North American Megadrought in 2020–2021," *Nature Climate Change* 12 (2022): 232–234; Cindy Yurth, "The Longest Drought," *Navajo Times*, September 5, 2011; "Livestock Purchase Program," *Navajo Times*, November 8, 1961.
30 Hosteenah Begay, Oral History, Chinle Agency, Envelope 149, Office of Navajo Economic Opportunity Collection, Navajo Nation Library, Window Rock, AZ; Emma Chief, Oral History, Tuba City Agency, Envelope 125, Office of Navajo Economic Opportunity Collection, Navajo Nation Library, Window Rock, AZ. See also Gladys Reichard, *Dezba: Woman of the Desert* (J. H. Augustin Publishers, 1936), 13 (description of use of Angora in weaving). While *Dezba* is a work of fiction, it is reflective of the extensive experiences of anthropologist and linguist Gladys Reichard, who lived among the Navajo during the 1930s.
31 Roy Kady, interview by author, July 13, 2023.
32 *Stories of Traditional Navajo Life and Culture by Twenty-Two Navajo Men and Women* (Navajo Community College, 1977), 102.

Chapter 3: How Goat's Milk Became Healthy

1 See "The Swill Milk Nuisance," *New York Times*, June 8, 1858. For a vivid description of distillery dairy and milk adulteration practices; see also John Mullaly, *The Milk Trade in New York and Vicinity* (Fowler and Wells, 1858), particularly chapter 4; Great Future for the Milk Goat," *Chicago Tribune*, October 5, 1919 (epigraph).
2 Robert Hartley, *An Historical, Scientific and Practical Essay on Milk as an Article of Human Sustenance* (Jonathan Leavitt 1842), 108, 200. See also E. Melanie DuPuis, *Nature's Perfect Food: How Milk Became America's Drink* (New York University Press, 2002), which takes an in depth look at Hartley's work and the parallel rise of milk consumption in the United States.
3 "The Swill Milk Trade of New York and Brooklyn," *Frank Leslie's Illustrated Newspaper*, May 8, 1858, 359.

4   Thomas M. Daniel, "The History of Tuberculosis," *Respiratory Medicine* 100 (November 2006):1862–1870; Arthur Stanley Pease, "Some Remarks on the Diagnosis and Treatment of Tuberculosis in Antiquity," *Isis* 31, no. 2 (April 1940): 380–393.
5   "Consumption in America," *Atlantic Magazine*, January 1869.
6   Alan Olmstead and Paul W. Rhode, "An Impossible Undertaking: The Eradication of Bovine Tuberculosis in the United States," *Journal of Economic History* 64, no. 3 (February 2004): 4–8; Pease, "Some Remarks on the Diagnosis and Treatment of Tuberculosis in Antiquity," 380–393 (tuberculosis-like disease in cattle described by Columella of Ancient Rome); Mitchell Palmer and W. Ray Waters, "Bovine Tuberculosis and the Establishment of an Eradication Program in the United States: Role of Veterinarians," *Veterinary Medicine International* (2011): 1–12. On the transmissibility of bovine tuberculosis see, for example, Mazyck P. Ravenel, "Tuberculosis and the Milk Supply," *Journal of Comparative Medicine and Veterinary Archives* 18, no. 12 (December 1897): 753–761, and Barbara Gutmann Rosenkrantz, "The Trouble with Bovine Tuberculosis," *Bulletin of the History of Medicine* 59, no. 2 (Summer 1985): 155–175.
7   "Death Lurks in Milk," *Nashville Banner*, Jan 30, 1897.
8   *Tuberculosis*, USDA Farmers Bulletin 473 (Government Printing Office, 1911), 8; Olmstead and Rhode, "An Impossible Undertaking," 31–32. To this day, public health officials continue to find tuberculosis infections in cattle populations of the United States, Canada, and Europe. In 2005, reports emerged that at least thirty-five New York City residents had contracted tuberculosis from raw milk cheese over a period of four years. See Marc Santora, "Tuberculosis Cases Prompt Warning on Raw Milk Cheese," *New York Times*, March 16, 2005.
9   I. Burney Yeo, "Clinical Lecture on the Contagiousness of Pulmonary Consumption Delivered in King's College Hospital," *British Medical Journal*, June 17, 1882, 895.
10 Nocard's data was widely cited during the period; see, for example, "An Invaluable Milk Supply," *Journal of the American Medical Association*, March 5, 1892; Dr. Carl G. Wilson, "Tuberculosis and the Milk Goat as a Valuable Aid in Preventing and Curing the Disease," *The Goat World*, October 1919, 7.
11 Louise Lippincott and Andreas Bluhm, *Fierce Friends: Artists and Animals, 1750–1900* (Merrell, 2005), 138–139. Adler's original painting Transfusion de sang de chèvre (Transfusion of Goat's Blood) is on display at the Museum of the History of Medicine in Paris; "Treatment of Tuberculosis by Goat's Serum," *Lancet*, August 24, 1901, 542; "New Cure for Consumption," *New York Times*, January 19, 1891.
12 "Intravenous Injection of Milk," *British Medical Journal*, August 28, 1880; on the practice of transfusing milk, see H. A. Oberman, "Early History of Blood Substitutes: Transfusions of Milk," *Transfusion* 9, no. 2 (March 1969): 74–77.
13 J. Finley Bell, "Some Fat Problems and Goat's Milk in Infant Feeding," *Archives of Pediatrics* 23, 204; "Goat's Milk," *British Medical Journal*, June 17, 1908. The latter article goes on to ask why more are not turning to goats as a solution to the problem of contaminated cow's milk. See also M. J. Rosenau, *The Milk Question* (Houghton Mifflin, 1912), 50.
14 Aharona Glatman-Freedman and Arturo Casadevall, "Serum Therapy for Tuberculosis Revisited: Reappraisal of the Role of Antibody-Mediated Immunity against Mycobacterium Tuberculosis," *Clinical Microbiology Review* 11, no. 3 (July

1996): 514–532 (summarizing serum research); S. H. McNutt and Paul Perwin, "Tuberculosis of Goats," *Journal of the American Veterinary Medical Association* 12, no.1 (April 1921): 82–84.

The potential healing properties of goat's blood has continued to draw interest from the medical community well into the twentieth century. During the 1990s, Gary R. Davis created a serum from goat's blood, which he believed could cure AIDS. The story of Dr. Davis was detailed in a podcast called *Serum* produced by WHYY in Philadelphia in 2022; see also Lisa Frazier, "The Goat Doctor," *Washington Post*, April 9, 2000.

15 The contaminated milk issues of the period caused some to look beyond animal milk entirely. In 1921, Henry Ford (inventor of the horseless carriage) suggested that an era of "cowless milk" was not far off; researchers in Massachusetts soon announced the development of a milk-type product concocted from oats, peanuts, and salt; see "Ford's Cowless Milk Shocks Even Chemists," *New York Tribune*, February 10, 1921; "Boston Experts Claim to Make Cowless Milk," *New York Times*, March 14, 1921. Around the same time, German researchers were experimenting with a milk substitute made from soybeans, see William Shurtleff and Akiko Ayoagi, *Early History of Soybeans and Soy Foods, 1900–1914* (Soy Info Center, 2021), available online at soyinfocenter.com.

16 "Goats Misunderstood," *Brooklyn Daily Eagle*, October 28, 1888; "American Goat: A Ridiculed Hero," *The Sunday Star* (Washington, DC), June 15, 1911.

17 Katherine Ott, *Fevered Lives: Tuberculosis in American Culture Since 1870* (Harvard University Press, 1996), 18–19. The words sanitarium and sanitorium were used interchangeably during this period as were other spelling variations such as sanatarium or sanatorium. I've used the term "sanitarium" to refer to the general concept of health or medical facility engaged in some form of tuberculosis treatment; when referring to specific facilities I've applied the term used by the facility.

18 See, for example, Michele A. Riva, "From Milk to Rifampicin and Back Again: History of Failures and Successes in the Treatment for Tuberculosis," *The Journal of Antibiotics* 67 (2014): 661–662, on the use of milk in Ancient Greece and Rome. Contemporary research continues to make connections between milk and the possible enhancement of antibiotic therapies for tuberculosis. See Virginia Meikle, Ann-Kristin Mossberg, Avishek Mitra, et al., "A Protein Complex from Human Milk Enhances the Activity of Antibiotics and Drugs against Mycobacterium Tuberculosis," *Antimicrobial Agents and Chemotherapy* 63, no. 2 (February 2019): .aac.01846–18.

19 "Last Minute Action Saves County Goats," *Chicago Tribune*, March 2, 1920; see also "Meeting favors Plan of County for Milch Goats," *Chicago Tribune*, January 27, 1920; Official Proceedings of the Board of Commissioners of Cook County, Illinois for the Year 1920–21, 67, 612, 852 (Goat Commission activities); "County Crowns Nanny Goat as Health Queen," *Chicago Tribune*, November 8, 1919. The Cook County Goat Commission appears to have been later absorbed into the Public Service Commission, which handled the administration of business related to the goats at Oak Forest starting in 1922.

20 "Mrs. James Patten's Goat on Strike or Lonesome," *Chicago Tribune*, April 17, 1920; "Mrs. J. A. Patten Sells Goat's Milk," *Chicago Tribune*, March 14, 1922; "Mrs. Patten Heads Goat Milk Association," *Chicago Daily Tribune*, December 7, 1932.

21 "Using Goat's Milk in Tuberculosis," *Philadelphia Enquirer*, Feb 18, 1906; see also "The Radnor Wayne Sanitorium for Tuberculosis," Hahnemannian Monthly News and Advertiser, March 1906, 34; "Sanitarium Gives Goat Milk Unqualified Endorsement," *Angora and Goat Milk Journal*, August 1921, 38; William Secor, "The Modern Milk Goat in Medicinal Practice," *Practical Medicine and Surgery*, July 1922, 14; *Prairie Past and Mountain Memories: A History of Dunseith, N. Dak., 1882–1982* (no publisher identified, [1982]); C. A. Higgins, *Las Vegas Hot Springs and Vicinity* (Passenger Department, Santa Fe Route, 1897), 26. A photo of the Toggenburg goats at the North Dakota Sanitarium appeared in the July 1935 issue of *Dairy Goat Journal*.

22 Sheila M. Rothman, *Living in the Shadow of Death: Tuberculosis and the Social Experience of Illness in American History* (Johns Hopkins University Press, 1994), chapter 9, beginning at 131.

23 "News Notes of the Southland," *Long Beach Telegram and Long Beach Daily News*, April 15, 1914; *The Bulletin of the Los Angeles County Medical Association*, July 20, 1922, 12; multiple ads for Dr. Pike's Health Resort in Long Beach (California) Press; see, for example, September 22, 1921; *The Goat World*, December 1924, 13. San Francisco General Hospital purchased goat's milk from Las Cabritas goat Ranch in Montara, California; see *Journal of Proceedings Board of Supervisors City and County of San Francisco* (Recorder Printing and Publishing Co., 1920), 820.

24 Harvey W. Wiley, "The Eternal Infantile," *Good Housekeeping*, June 1916, 201.

25 Physicians Department, *American Standard Milch Goat Keeper*, December 1914, 254; "A Practical Study of Goat's Milk in Infant Feeding as Compared to Cow's Milk," *The American Journal of Obstetrics and Diseases of Women and Children* (June 1913): 1245; L. E. Bonsieur, "The Modern Milk Goats," *The Goat World*, August 1922, 92; "Goats Milk to Get Test," *The Modern Hospital*, March 1915, 232.

26 W. H. Jordan and G. A. Smith, "Goats Milk For Infant Feeding," *New York Agricultural Experiment Station Bulletin* 429, February 1917.

27 Vivian Wiser, Larry Mark, H. Graham Purchase, eds., *100 Years of Animal Health 1884–1984* (Associates of the National Agricultural Library, 1987), which compiles articles from *Journal of NAL Associates* 11, nos. 1–4 (January–December 1986). The Bureau of Animal Industry was eliminated in 1953 as part of a reorganization of the USDA.

28 Christina Reh Wyse, "William Black and the Southwestern Livestock Industry" (master's thesis, Texas Tech University, 1995), 84–85, William L. Black, *Complaint of William L. Black of Ft. McKavett, Texas Against US Department of Agriculture* (USDA, 1926); George F. Thompson, *The Angora Goat*, Farmer's Bulletin 127 (Government Printing Office 1901).

29 Accounts of the circumstances surrounding Thompson's death appear in a variety of publications including Wesley William Spink, *The Nature of Brucellosis* (University of Minnesota Press, 1956), 12; John R. Mohler and George H. Hart, "Malta Fever and the Maltese Goat Importation," in *United States Department of Agriculture Annual Report of the Bureau of Animal Industry for the Year 1908* (Government Printing Office, 1910); "Goats for Uncle Sam," *New York Times*, June 24, 1905.

30 S. R. Winters, "Uncle Sam's Goat Dairy," *The Goat World*, May 1922, 15; C. G. Potts, "Government Experiments in Goat Improvement," *The Goat World*, July 1925, 1. See also the children's book *Uncle Sam's Animals* by Frances Margaret

Fox (The Century Co., 1927). In the chapter "Uncle Sam's Little Kids and Little Lambs," Fox mentions the varieties of sheep and goats kept on the experimental farm, noting that "Uncle Sam experiments with goats, that babies everywhere may have more and better milk."

31 On recent research regarding goats and tuberculosis, see for example: Javier Bezos, Lucía de Juan, Beatriz Romero, et al., "Experimental Infection with *Mycobacterium Caprae* in Goats and Evaluation of Immunological Status in Tuberculosis and Paratuberculosis Co-Infected Animals," *Veterinary Immunology and Immunopathology* 133, no. 2–4 (February 2010): 269–275; Sabrina Rodríguez, Javier Bezos, Beatriz Romero, et al., "*Mycobacterium Caprae* Infections in Livestock and Wildlife, Spain," *Emerging Infectious Diseases* 17, no. 3 (March 2011): 532–535.

Chapter 4: The Goat's Milk Business

1 Mrs. John A. Logan, *The Part Taken by Women in American History* (Perry-Nalle Publishing Co., 1912), 375; "Stevens Sale of Registered Goats Planned," *Lake Geneva Regional News*, June 28, 1923 (epigraph).
2 Elizabeth Higgins, "The Fatality of Whiskers," *Chicago Tribune*, January 24, 1909; George Fayette Thompson, *Information Concerning the Milch Goats* (Government Printing Office, 1905), 51.
3 Thompson, *Information Concerning the Milch Goats*, 48.
4 According to one account, Shafor imported the goats on behalf of someone else, who asked him to do so because Shafor had prior experience importing sheep. Because of the record of the import, Shafor got on the radar of the USDA, and George Thompson contacted Shafor about the animals. Shafor reluctantly became involved with the AMGRA board because he had experience as a member of the Oxford Down Sheep Association board. The first AMGRA dairy goat registrations were entered on Oxford Down Sheep Association forms. Shafor stayed on with AMGRA at the insistence of Mrs. Roby (information by former ADGA historian Shari Reyna, collection of author).
5 "National Goat Farm Helps Sick Babies," *The Marshall Herald*, April 11, 1924.
6 *Wool Markets and Sheep*, November 1904, 19.
7 Thompson, *Information Concerning the Milch Goats*, 49 (fair exhibits); "One Boy's Success with Angoras," *The Shepherd's Criterion*, July 1905 (reprint of Cohill article from *The American Boy*). According to some accounts, two Schwarzwald Alpine goats imported from Germany were featured at the St. Louis Fair in a display called Carl Hagenbeck's Animal Show. Hagenbeck was a German businessman who traded in all manner of wild animals and is often credited as a developer of the modern zoo. Although there were only a few dairy goats on display, Angora goat breeders brought 300 animals to the fair, and George Thompson served as the official judge of the Angora Goat Show.
8 "Goat No Longer a Joke Says Mr. Paine of Franklin," *Boston Globe*, May 15, 1921.
9 "One Family's Solution of the Milk Problem," *Suburban Life*, October 1909, 183. An earlier, substantially similar article by Bull was published in *Good Housekeeping*, April 1908, titled "Our Backyard Dairy." Bull published articles about goats and goat keeping in a number of periodicals of the day including *Country Life in America* magazine. See also his book on the topic, *Money in Goats* (Wakefield Co. 1911).

10 Filomena Gould, "Information Plus," *Indianapolis News*, January 16, 1945; "Milk Bar is Feature at Swiss Goat Dairy," *Indianapolis News*, May 30, 1941.

11 "Scientific Methods and Sanitation Help in Profitable Production of Goat Milk," *Rochester Democrat Chronicle*, August 8, 1923 (Haedtler), see also "Goat Farm Owner is "Good Angel" for Many Unfortunates," *Southtown Economist* (Chicago), July 28, 1925; "Pitts Goat Dairy Supplies Rich, Easily Digestible Goat's Milk," *Atlanta Constitution*, March 19, 1951; *The Goat World*, April 1925, 7 (Denver goat dairies); Press Huddleston, "Pioneer Goat Dairyman Serves City," *Atlanta Journal Constitution*, May 1, 1950 (Pitts).

12 Irma B. Mathews, "The Milch Goat in Southern California," *American Sheep Breeder and Wool Grower*, February 1916, 112. A photo of the 1916 float graces the cover of the first issue of *The Goat World*, published in February 1916. California goat farmers represented 35 percent of the overall membership of the American Milch Goat Record Association, followed by second-place Ohio with 21 percent. See *The American Milch Goat Record*, vol. 1 (United Brethren Publishing House, 1914). In the first three volumes, the organization helpfully lists its membership alphabetically by state.

13 "Goats Going North," *Oregonian*, August 6, 1917; "Islands to be Nation's Goat Center," *Bellingham Herald*, August 14, 1917; "Feed Fit for King," *Oregonian*, June 26, 1918; "She'll Be Goat Herder," *Seattle Star*, March 6, 1924; "Sanitarum on a Goat Farm," *Northwest Medicine*, October 1919, 214; *Angora and Goat Milk Journal*, July 1920, 34; "Seeks Some Cash," *Bellingham Herald*, April 23, 1924. According to news reports, for a time the island dairy sent some of its milk to a goat cheese factory in nearby Stanwood, Washington, on the mainland.

14 "Stevens Sale of Registered Goats Planned," *Lake Geneva Regional News*, June 28, 1923. Agawam is a reference to an Indigenous tribe of coastal New England, where Stevens's family originated.

15 Howard Kegley, "Southland's Prize Goats to Stock Chicago Millionares' Dairies," *Los Angeles Times*, April 21, 1922; "California Goats Sold at Auction," *The Goat World*, July 1922, 13; "Pure Bred Goat Makes Bow at Agawam Sale," *Chicago Tribune*, June 25, 1922.

16 For background on Meyenberg and the development of the condensation industry generally, see L. L. De Bra, "California's Condensed Milk Industry," *Pacific Dairy Review*, July 8, 1915, cover story, and "The Story of Condensed and Evaporated Milk," *Pacific Dairy Review*, July 15, 1915, 608.

17 "Getting Goat Milk into Cans," *Pacific Dairy Review*, December 20, 1917; "Raising Goats New Industry," *Santa Cruz Evening News*, February 22, 1916; *The Californian*, May 4, 1916. Details about the partnership from Corporation Records, California Secretary of State.

18 "Chat," *King City Rustler*, May 10, 1918; "Goat Milk Co's Big Contract," *The Californian*, May 5, 1917; "Widemann G. M. Co. Moves to Pescadero," *King City Rustler*, January 11, 1918. The Widemann corporation was suspended in 1926.

19 "Sons Carry on What John B. Meyenberg Started," *The Goat World*, September 1928, 17; "Meyenberg Will Can Goat Milk in Paso Robles," *King City Rustler*, April 6, 1936.

20 "Goat Farming: A New Industry for Women," *American Standard Milch Goat Keeper*, March 1915, 329; Alice Brown "Goat Dairying, a Vocation for Women," *The Goat World*, January 1922, 30; "The Raising of Goats as a Business for Women," *Los Angeles Times*, May 16, 1920. See also Mrs. D. L. Bunnell,

California's Women Goat Ranchers," *University of California Journal of Agriculture*, February 1920.

21 "Goat Raising Stanford Girl's Latest Success," *San Francisco Examiner*, April 23, 1916; Irmagarde Richards, *Modern Milk Goats* (J. B. Lippincott Co., 1921), 70, 72.

22 "Here's World's Only Woman Goat Buyer," *San Francisco Chronicle*, March 28, 1923; "California Milk Goats Bound for the Argentine," *Los Angeles Times*, May 6, 1923; "Woman to Take Goats on Voyage," *Los Angeles Daily Times*, March 28, 1923; "Queen of the Goat Buyers," *Oakland Tribune*, July 22, 1923.

23 "Enthusiastic Over University Goat Herd," *California Aggie*, October 15, 1924.

24 Frances Duncan, "They Tried Chickens but Found it Was More Fun to Raise Goats," *Los Angeles Times*, April 24, 1927; "Cultured Women Make Goats Pay," *Los Angeles Times*, February 7, 1927.

25 "Humboldt County Valleys Filling with Goat Herds," *Santa Rosa Press Democrat*, October 7, 1922; "Southern Humboldt has Promising New Industry," *Humboldt Times*, September 10, 1922 (article contains several photos); "Rapid Progress of Goat Industry Marked by Success of Cooperative Cheese Factory," *Humboldt Times*, August 2, 1925 (more photos). See also "Humboldt to Produce Goat Cheese," *Pacific Dairy Review*, June 22, 1922.

26 "Goat's Milk Instead of Cow's Milk," *Camden Post-Telegram*, August 31, 1911 (Palisades Park); "Goat Herds to Solve Poor's Milk Problem," *Minneapolis Star Tribune*, September 14, 1911; E. I. Farrington, "Milk Goats in New Hampshire," *Rural New Yorker*, March 15, 1919, 462; Elmer F. Dwyer, "A Visit to a Grecian's Goat Farm," *American Standard Milch Goat Keeper*, January 1916, 5; "Greek Cheese Made in New Hampshire," *Boston Sunday Globe*, December 19, 1915.

27 See, for example: "Castle Gate," *Eastern Utah Advocate*, April 3, 1902 (goat importations); "Cheese Factory Found Unsanitary," *Salt Lake Tribune*, August 13, 1914; Cretans Hidden High Among Hills of Utah are Making Goat Cheese," *The Pomona Progress*, May 1, 1915, "Cheese of Goat's Milk Made Here," *Salt Lake Herald Republican*, May 26, 1918; Philip Notarianni, "Italianata in Utah: The Immigrant Experience," in *The Peoples of Utah*, ed. Helen Z. Papanikolas (Utah State Historical Society, 1976). Luigi Nicoletti's son Tony continued the family cheesemaking tradition at Nicoletti Cheese House in Salt Lake City during the 1960s and '70s. According to one newspaper, goat rancher Sevenstad Larson was marketing goat cheese as early as 1897 in Laramie, Wyoming; see "From Our Exchanges," *Deseret News*, September 15, 1897.

28 Home Department Section, *San Bernadino Sun*, April 8, 1922. The article notes that goat meat from the farm is also shipped to Italian communities in Los Angeles. "Milch Goat Farms Center in Hills of El Dorado County," *Sacramento Bee*, May 17, 1919 (El Dorado County);"Protests Making of Goat's Milk Cheese," *Sacramento Daily Union*, June 9, 1914; On Andreolli, see T. C. Holt, "Goats Helping to Solve National Forest Fire Problem," *Los Angeles Times*, November 27, 1921, and "Wanted: More Goat Ranches in the Higher Foothills Areas," *Los Angeles Times*, April 26, 1924.

29 Unpublished memoir, Edwin C. Voorhies Papers, D-089, Department of Special Collections, General Library, University of California, Davis. Although Voorhies spent nearly a decade studying goats, his memoir devotes only one paragraph to the topic.

30 "Roquefort Now Being Made at UC Farm Plant," *Oakland Tribune*, February 3, 1924; S. A. Hall and C. A. Phillips, *Manufacture of Roquefort Cheese From Goat's Milk*, University of California College of Agriculture Bulletin 397 (University of California Printing Office, 1925).

31 On the Bransons see, for example., Alice Latta, "Oregon's First Roquefort Cheese Made in Mountains Above Falls City," *Sunday Oregonian*, October 4, 1925; Lillie L. Madsen, "Roquefort Cheese Factory Thrives in Green Covered Hills Near Falls City," *Sunday Oregonian*, September 11, 1927; "The Milk Goat Industry a Comer," *Oregon Statesman*, July 27, 1922. Fannie Branson later moved to the Oregon coast and became known for carving miniature horses. See "Hobbyist Makes Tiny Horses of Balsa Wood and Calfskin," *Popular Mechanics*, January 1946, 49. On Montchalin: "Montchalin Wins Government Approval," *Angora and Goat Milk Journal*, April 1921, 33. See also "Marshall Revels in Glories in Mountain Springtime Beauty," *Siskiyou Daily News*, April 3, 1924 (Ettersburg); "Races Held on Holiday at Capital," *The Californian*, September 7, 1936 (Mount Lassen). To the north, on Salt Spring Island in British Columbia, Colonel Jasper Bryant also made a roquefort-style cheese from the milk of his herd of goats during the 1920s.

Chapter 5: Back to the Land

1 Editorial, *The Goat World*, January 1945, 14; Irmagarde Richards, "What's Wrong with the Milk Goat Industry," *The Goat World*, June 1922, 10 (the series continues in the July issue). The 1945 editorial cites turmoil within the ranks of goat breeders as one of the main reasons for the failure to progress as an industry. For years, members were wrapped up in discussions regarding breeding and registration rules that became so heated that several offshoot registry organizations were formed, including the American Goat Society; Bill Morgan, ed., *Selected Letters of Allen Ginsberg and Gary Snyder, 1956-1991* (Berkeley, CA: Counterpoint 2009), 104 (epigraph).

2 See, for example, Margaret E. Dean, "Goat Milk and the War," *The Goat World*, June 1942; M. W. King, "Let's Win this War," in *American Dairy Goat Yearbook* (C. W. Romer, 1942); I. B. Boughton, "Keep Your Milk Goats Producing for Victory," *The Goat World*, July 1944. Goat periodicals published during the 1940s are full of articles about wartime marketing strategies.

3 Jennifer Bice, interview by author, July 21, 2021; on Rodale, see Wade Green, "Guru of the Organic Food Cult," *New York Times*, June 6, 1971; Julia Trent, "Plenty of Kids in Her Busy Life," *The Press Democrat* (Santa Rosa), July 22, 1973.

4 Bill Kovach, "Communes Spread as the Young Reject Old Values," *New York Times*, December 17, 1970. See also Timothy Miller, *The 60s Communes: Hippies and Beyond* (Syracuse University Press, 1999).

5 The Houriet quotation is related by Lucy Horton in *Country Commune Cooking* (New York: Coward, McCann and Goeghan Inc., 1972) 13–14. Horton was Houriet's typist when he drafted *Getting Back Together*.

6 "Voluntary Primitivism," in Unohoo, Coytoe, Rick, and the Mighty Avengers, *Morningstar Scrapbook* (Friends of Morning Star, 1976), 159; Stephen Gaskin, *Hey Beatnik: This is the Farm Book* (The Book Publishing Co., 1974), unpaginated.

7 Nancy Bubel, "Notes by a New Goat Keeper," *Organic Gardening and Farming*, March 1971, 104–111.

8 Nancy Pierson Ferris, "Get Your Goat," *Mother Earth News*, no. 6, 1970.

9   Lucy Horton, interview by author, October 24, 2022; Robert Houriet, *Getting Back Together* (Avon Press, 1971), 142.
10 Gordon Ball, personal correspondence with author; Judith Margolis, interview by author, October 18, 2022; the same events are also covered in David Margolis, *Change of Partners* (Permanent Press, 1997), and Elaine Sundancer, *Celery Wine: the Story of a Country Commune* (Community Publications Cooperative, 1973).
11 Ramon Sender Barayon, ed., *Home Free Home: A History of Two Open Door California Communes* (Calm Unity Press, 2017), 26; Raymond Mungo, *Total Loss Farm: a Year in the Life* (Dark Coast Press Pharios Editions, 2014), "The Kindly Goat Lady of Westhaven," *The Times-Standard* (Eureka, CA), January 24, 1971.
12 Mary Keehn, interview by author, July 22, 2021, Kate Coleman, "Country Women: The Feminists of Albion Ridge," *Mother Jones*, April 1978, 23; Mildred Hamilton, "Farming as a Way of Life for Women from the City," *San Francisco Examiner*, September 5, 1976; "Barn Building, Fence Mending, Goat Raising Well Digging Women,*" Ms.*, August 1974, 22. Sherry Thomas and Jeanne Tetrault, *Country Women: A Handbook for the New Farmer* (Anchor Press, 1976), xv. See also Dona Brown, *Back to the Land: The Enduring Dream of Self Sufficiency in Modern America* (University of Wisconsin Press, 2011).
13 C. E. Leach, "Conclusions," *Dairy Goat Journal*, February 1960, 23.
14 This discussion is driven in part by Charlotte Biltekoff, *Eating Right in America: The Cultural Politics of Food and Health* (Duke University Press, 2013).
15 Gypsy Boots, *Bare Feet and Good Things to Eat* (Virg Nover, 1965), 124. In a 1974 episode of *What's My Line*, another popular television show, goat farmer Jospehine Eberhardt stumped the panelists with her occupation (goat farmer)— though no actual goats made an appearance. See also Jonathan Kaufman, *Hippie Food* (Harper Collins, 2018).
16 Bernard Jensen, *Goat's Milk Magic* (Bernard Jensen, 1994); "Dude Ranch Sold to Doctor for Health Retreat," *Times-Advocate* (Escondido, California), March 16, 1954. After changing hands several times, Briar Hills Dairy exists today as Mt. Capra Products, still headquartered in Chehalis, Washington.
17 Margaret Kilgore, "Average Health Food User Isn't 'Nut,'" *Los Angeles Times*, April 14, 1975; "Food Co-Op Offers Discounts for Work," *The Berkeley Gazette*, November 13, 1979; Jessie Bell, "Health Food Movement Has Come a Long Way," *Los Angeles Times*, January 23, 1972; Jean Hewitt, "Buying Health foods is Easier Now," *New York Times*, June 19, 1971.
18 "Dairy Goat Business is Booming," *Sonoma West Times and News*, July 13, 1972; Barbra Sullivan, "Counterculture to Mainstream: Health Food's Journey into the 90s," *Chicago Tribune* June 2, 1988.
19 "Aim to Resurrect Crawford County with 'Design for Living' Plan," *Boscobel Dial*, August 28, 1969.
20 Michael Hankin, interview by author, May 27, 2022; Kathleen Piper, interview by author, October 26, 2021. See also "Kickapoo to You Too," *Cheesemakers Journal*, April–May 1982 (published by the New England Cheesemaking Co.). There must have been something in the western Wisconsin air that spawned cooperatives; during the 1980s the Organic Valley Cooperative, now a national force in the cow dairy industry, started in La Farge, Wisconsin, just up the road.
21 The Southwestern cooperative was not the only goat dairy cooperative in the state. In Burnett County, in northern Wisconsin, another small group of goat dairy farmers formed the Dairy Goat Products Cooperative.

22 Al Swegle, "Make Goat Cheese," *The Gazette* (Cedar Rapids, IA), September 14, 1969; see also Kauffman, *Hippie Food*, chapter 7, for a detailed discussion of consumer co-ops in the United States.

23 "School Job: Dairy Goats: Cheese," *Corvallis Gazette-Times*, August 25, 1971; Cam Montgomery, "Members of Goat Co-op Aren't Kidding Around," *The Capital Journal*, October 30, 1971; Judy Kapture, "Dairy Goat Industry is Expanding," *Animal Industry Today* (1978), 2.

24 Beryle E. Stanton, "Co-ops Pull Together to Can Goat Milk," *News for Farmer Cooperatives*, March 1951, 11; according to one source I spoke with, the co-op formed when Meyenberg stopped buying milk from some local producers; the effort to create an independent co-op was an effort to ensure their own continuing livelihoods. See also "Marketing Considerations of Goat Milk," Seventh Annual California Dairy Goat Day, University of California, Davis, November 7, 1981, Department of Special Collections, General Library, University of California, Davis.

25 "Goat Dairymen Sue Over Plant's Closure," *Modesto Bee*, September 5, 1976.

26 "Goat Business is One of a Kind Operation," *Longview News Journal*, Dec 27, 1987; Joe Bigham, "Goat Milk Bucks Ag Trend," *Fresno Bee*, March 22, 1987; Frank Fillman, interview by author, August 7, 2023; John Jeter, interview by author, May 26, 2023; Tony Walker, "Goat Dairymen Construct Plant," *Modesto Bee*, October 20, 1976; "Goat Business is One of a Kind Operation," *Longview News Journal*, Dec 27, 1987; Joe Bigham, "Goat Milk Bucks Ag Trend," *Fresno Bee*, March 22, 1987; Leo Dollar, "Dairy Goat Herds on Upswing in Central Valley," *Modesto Bee*, February 9, 1980.

27 "Jackson-Mitchell Co. Purchases Ozark Firm," *Dairy Goat Journal*, April 1977, 28. According to John Jeter, former CGDA general manager, for a time the CGDA had considered purchasing Jackson-Mitchell. A FOIA request for US Department of Justice documents related to the anti-trust case was pending at the time of this book's publication.

28 David Gumpert, "Still Another Index Confirms Recession: The Goat Indicator," *Wall Street Journal*, January 16, 1975; Richard Orr, "Milk Costs Got Your Goat? Get it Back!," *Chicago Tribune*, December 19, 1975; "Two Sets of Quadruplets Equals Eight Kids," *Santa Ynez Valley News*, March 9, 1972; "DHIA reports Goat Boom, New High for Cows," *Fresno Bee*, June 20, 1976. Unfortunately USDA Agricultural Census figures did not separate out goat population figures by industry until the 1980s, so it is difficult to determine specific dairy goat population numbers during this period

29 See, for example, John D. Faulkner, Darold W. Taylor, and Irving H. Schlafman, "The USPHS Method of Rating Milk Supplies and Its USE in the Interstate Milk Shipper Program," *Journal of Food Protection* 25, no.9 (September 1962): 277–281. It's worth noting that pasteurization was controversial since the earliest years of its implementation in the late nineteenth century because of concerns about the effect of heat treatment on milk's inherent nutritional qualities.

30 United States Public Health Service, *What You Should Know About Grade A Milk*, Public Health Service Publication 1472 (US Department of Health, Education, and Welfare, 1966); Jean Mayer, "Raw Milk Has No Unusual Nutrients," *New York Daily News*, December 13, 1972 (Mayer's column ran in newspapers across the country).

31 "Plea Renewed to Ease Ban on Goat Milk," *Daily Oklahoman*, January 7, 1960; "Council Eases Ruling on Sale of Goat's Milk," *Daily Oklahoman*, January 27,

1960; "Can't Sell Goat Milk, She Gives it Free," *Capital Times*, March 29, 1975; Susan Schwarz, "Redmond Goat Dairy Began with Modern Day Heidi's Little Pet," *Seattle Times*, November 23, 1969. Mystic Lakes Goat Dairy was also facing zoning changes in Redmond, Washington, which would eventually push the dairy out of business. For Kentucky laws, see Unpasteurized Goat Milk, 902 KAR 50:120; "Bootleggers Causing Crisis for Sate's Lone Legal Goat Milk Operation," *The Roanoke Times*, October 27, 1979.

32. Stephen Ferris, "Goat Farm Operator Turns Diarist for National Farm Magazine," *Modesto Bee*, July 21, 1979; Jack Hawes, interview by author, September 17, 2021; "New Laurelwood Acres Goat Dairy Will Hold Open House This Weekend to Celebrate Start of Operations," *Ripon Record*, June 23, 1966.

33. "The Simple Pleasures of Breeding Dairy Goats," *Sonoma West Times and News*, April 20, 1983; "Couple Isn't Kidding Around," *The Press Democrat*, November 9, 1987; author interview with Jennifer Bice.

Chapter 6: Say Chevre

1. "Chevre: a Creamy French Style Goat Cheese Now Produced in California," *San Francisco Examiner*, September 30, 1981; Marian Burros, "California Cuisine: Assessing its State," *New York Times*, March 3, 1982 (epigraph).

2. Caroline Bates, Gourmet, October 1975; James Beard, "A Dinner of Surprises," *San Francisco Examiner*, April 5, 1978; Beverley Stephen, "American Chefs are the New Rising Stars," *New York Daily News*, July 5, 1979. Waters discusses the New York dinners in her memoir *Coming to My Senses: The Making of a Counterculture Cook* (Clarkson Potter, 2017).

3. "This Couple Gave All for Cheese," *Contra Costa Times*, May 13, 1977; Charles L. Sullivan Papers on California Wine History, D-346, Department of Special Collections, General Library, University of California, Davis (among this collection are monthly newsletters from the Wine and Cheese Center).

4. "Chenel, Laura" (interview), June 1, 2010, Joyce Goldstein Papers on California Cuisine, Department of Special Collections, General Library, University of California, Davis; Joyce Goldstein, interview by author, July 29, 2021.

5. Janet Fletcher, interview by author, October 31, 2019; Author Interview with Joyce Goldstein; Thomas McNamee, *Alice Waters and Chez Panisse* (Penguin Press, 2007), 160; Gael Green, "Gael Grazes," *New York Magazine*, September 30, 1985; Mimi Sheraton, "Books: I Cook, Therefore I Am," *Time*, November 24, 1986.

6. Craig Arnoff and John L. Ward, ed., *Contemporary Entrepreneurs* (Omnigraphics, 1992), 91.

7. Eunice Fried, "The Triumph of Chevre," *The Monthly Magazine of Food & Wine*, April 1981, 36 (the cover proclaims "Chevre: Cheese of the Year"); Harvey Steinman, "Laurie Chenel, a Nanny to a Herd of Goats, Makes Her Whey into the Cheese Business," *People Weekly*, July 12, 1982, 90; Harvey Steinman, "For Royal Menus, a Definite California Flavor," *San Francisco Examiner*, March 2, 1983; Ruth Reichl, "The Making of an Entrepreneur," *Working Woman*, August 1984, 95.

8. Steve Jenkins, *Steve Jenkins Cheese Primer* (Workman Publishing, 1996). A full list of all of Jenkins' American Treasures is on 384. Rainbow Chevre's Berkshire Blue predated a later cheese also named Berkshire Blue, made from Jersey cow's milk, which was produced in Massachusetts in the 2000s.

9. Nancy Masumoto, "The Nanny Diaries: From Coach Leather to Coach Farm," *Edible Manhattan*, March–April 2012. Cahn also detailed his exploits starting and running a goat dairy in the book *The Perils and Pleasures of Domesticating Goat Cheese: Portrait of a Hudson Valley Dairy Goat Farm* (Catskill Press, 2003).

10. Camilla Stege, interview by author, October 19, 2021; Penny Duncan, interview by author, August 23, 2021, Barbara Brooks, interview by author, October 8, 2021; Warren Wolfe, "Cheesemaker Looks to Goats," *Minneapolis Star Tribune*, September 28, 1975; Arlo Jacobsen, "New Popularity for Goats in Iowa," *Des Moines Register*, June 27, 1976.

11. Judy Schad, interview by author, August 23, 2021; Christine and Vincent Maefsky, interview by author, May 27, 2022; Suan Matthis-Johnson, "Looking Back at West Virginia's Back to the Landers," *Daily Mail WV*, April 18, 2019.

12. Jean Claude Le Jaouen's book was later translated into English and sold by the New England Cheesemaking Supply Company as *The Fabrication of Farmstead Goat Cheese* (Cheesemakers' Journal, 1990); Mike Moore, "An Organic Chevre as a Labor of Love," *New York Times*, May 23, 1990.

13. Ricki Carroll, interview by author, January 14, 2020.

14. Miriam Ungerer, "These Pots and Pans Play Leading Roles," *New York Daily News*, April 23, 1978; Suzanne Hamlin, "National Archive For Gastronomes," *New York Daily News*, September 29, 1982, Author interview with Jennifer Bice, Clark Wolf, interview by author, April 5, 2022.

15. Evan Jones, "New American Cheesemakers," *Gourmet*, January 1985, 48; Author interview with Mary Keehn.

16. Bruce Naftaly, interview by author, October 25, 2021. See also "Small Pennsylvania Dairy Makes Profit," *United Caprine News*, September 1991, 31 (Newbold). On David Greatorex, see Roy Andries DeGroot, "Our Fine Little Cheeses That Equal Europe's," *Chicago Tribune*, March 23, 1981.

17. Author Interview with Judy Schad; Dahlia Ghabour, "Legendary Louisville Chef Announces Retirement and Will Close Restaurant in June," *Louisville Courier Journal*, June 2, 2020.

18. Emmi also acquired the San Francisco Bay Area's Cowgirl Creamery (maker of cow's milk cheeses) along with Tomales Bay Foods, Cowgirl Creamery's distribution arm, in 2016. Emmi had previously acquired Wisconsin-based Roth Kase, a cow's milk cheesemaker, in 2009. In 2021, Saputo also acquired a manufacturing facility in Reedsburg, Wisconsin, that produces goat milk whey.

19. United States Department of Agriculture, Agricultural Statistics Board, National Agricultural Statistics Service, "Sheep and Goats," January 31, 2025, https://downloads.usda.library.cornell.edu/usda-esmis/files/000000018/zk51xc07n/9593wq66x/shep0125.pdf; see also "Say Cheese: With $10M Investment, La Clare Creamery is Ready for its Close UP," *Wisconsin State Farmer*, October 15, 2019.

20. Emily Green, "The Goat Cheese Divas," *Los Angeles Times*, July 10, 2002. On frozen milk see, for example, "Using Frozen Dairy Goat Milk at Pitts Dairy in Atlanta, Georgia," *Dairy Goat Journal*, November 1961, 4, and "Deep Freezing Goat Milk," *Dairy Goat Journal*, June 1965, 3. As the Laura Chenel Company was rapidly expanding during the mid-1990s, the company began purchasing frozen curd from Norway because it couldn't produce enough milk to meet demand.

21. See, for example, Mary Beth Matzek, "Strong Dollar Hitting Goat Dairies Hard," *Wisconsin Dairy Farmer*, September 6, 2017.

22 Veronica Pedraza, interview by author, October 25, 2022; Mateo Kehler, interview by author, October 25, 2022.
23 Grace Garwood, "Growth Prospects Strong for Plant Based Cheese," foodinstitute.com/focus/growth-runway-significant-for-plant-based-cheese/; on the history of cheese alternatives, see William Shurtleff and Akiko Ayoagi, *History of Cheese, Cream Cheese and Sour Crean Alternatives* (Soyinfo Center 2013), available online at at Soyinfocenter.com.

Chapter 7: Urban Goats

1 Jennie Grant, interview by author, December 9, 2019; John Metcalfe, "Let's Goat Crazy," *Seattle Weekly*, September 12, 2007.
2 Jeninne Lee St. John, "Urban Animal Husbandry," *Time*, August 17, 2009; Jennifer Bleyer, "Fresh Goat Milk, Dead Wood and Dubious Neighbors," *New York Times*, February 22, 2011.
3 "Goat, Nuisance to Police for Days, Prisoner at Last," *Fort Worth Star-Telegram*, December 4, 1933; "Pigs Ruled Out of the City, Goats Go by Request," *Spokane Press*, May 21, 1936.
4 Rick Kogan, *A Chicago Tavern: A Goat, a Curse, and the American Dream* (Lake Claremont Press 2006), 20.
5 "Woman and Her Eighteen Goats Defy Law and Progress," *San Francisco Examiner*, April 28, 1951; "The Memories of the Goat Lady of Potrero Hill," *San Francisco Examiner*, May 5, 1982.
6 "Wandering Billy Gets Police Goat," *Sunday News*, November 7, 1971; *North East Bay Independent and Gazette*, July 17, 1979; Al J. Laukaitis, "Pygmy Goat Runs Afoul of Law," *Lincoln Journal Star*, November 1, 1995.
7 This discussion of urban goat keeping laws follows categories outlined by William Hale Butler in "Welcoming Animals Back to the City: Navigating Tensions of Urban Livestock Through Municipal Ordinances," *Journal of Agriculture, Food Systems and Community Development* 2, no. 2 (2012): 1–23; Nathan McClintock, Esperanza Pollanna, and Heather Wooten, "Urban Livestock Ownership, Management and Regulation in the United States: An Exploratory Survey and Research Agenda," *Urban Studies Planning Faculty Publications and Presentations* 90 (2014): 426–440.
8 Ads from *Los Angeles Express*, December 9, 1911, and February 29, 1912; Rachel Surls and Judith Gerber, *From Cows to Concrete: The Rise and Fall of Farming in Los Angeles* (Angel City Press 2016); Ann Simmons, "Owners Cling to Tiny Farms in LA," *Los Angeles Times*, January 23, 2012; Susan Carrier, "Where the Fast Lane Slows Down," *Los Angeles Times*, October 24, 2004.
9 On *pajarete*, see, for example, Bill Esparza, "Inside the Clandestine Milk Fueled Ranch Parties of Southern California," *Eater Los Angeles*, November 29, 2022, https://la.eater.com/2022/11/29/23483129/secret-mexican-ranch-parties-pajaretes-raw-milk-cocktails-los-angeles-california-feature. In 2014, neighbor complaints brought police to Richland Farms; police eventually cited a resident for "distributing illegal milk." See Samantha Shaefer, "Compton Resident Cited for Selling Raw Milk," *Los Angeles Times*, January 30, 2014.
10 "Brighton Gets Their Goats For Good," *Brighton Blade*, March 8, 2023, https://coloradocommunitymedia.com/2023/03/08/brighton-gets-their-goats-for-good/.

11 Steve Pardo and Christine Feretti, "Fight to Reclaim Goats takes aim at Detroit Law," *Detroit News,* Jan 3, 2015; David Sands, "Efforts to Update Detroit's Livestock Ordinance in Play Again," *Planet Detroit,* September 8, 2022, PlanetDetroit.org.
12 See, for example, Alexia Elejalde-Ruiz, "Chicago Proposal to Limit Chickens Raises Hackles," *Chicago Tribune,* October 15, 2019.
13 Jennifer Blecha, "Regulating Backyard Slaughter: Strategies and Gaps in Municipal Livestock Ordinances," *Journal of Agriculture, Food Systems and Community Development* 6, no. 1 (Fall 2015): 33–48; Ye Thian, "At Planning Meeting, Oaklanders Debate Over Urban Animal Husbandry," *Oakland North,* July 22, 2011, https://oaklandnorth.net/2011/07/22/at-planning-meeting-oaklanders-debate-over-urban-animal-husbandry/; Matthai Kuruvila, "Permit a Pest to Gardener Who Sells Food," *San Francisco Chronicle,* April 1, 2011; Tim Anderson, "Oakland Should not Allow Backyard Livestock," *Oakland Tribune,* October 17, 2012. Novella Carpenter details her experiences keeping goats on her urban farm in her second book, *Gone Feral: Tracking My Dad Through the Wild* (Penguin Press, 2014).
14 Miami residents' experiences with animal carcasses and body parts are well documented, see for example, Linda Robertson, "A Severed Cow's Tongue or Rooster Legs May Show Up on Your Running Path," *Miami Herald,* April 14, 2019; Church of the Lukumi Babalu Aue, Inc. vs. City of Hialeah, 508 US 520 (1993).
15 Gianaclis Caldwell, personal correspondence with author.
16 Lainey Morse, interview by author, November 7, 2019; Hope Hall, interview by author, October 17, 2024; Margaret Hathaway, interview by author, October 24, 2022.
17 Choi Chatterjee, interview by author, January 8, 2020; Karen Krivit, interview by author, May 25, 2021; Genevieve Church, interview by author, June 1, 2021. Tragically, the Chatterjee-Sayeed house was destroyed, along with much of their Altadena neighborhood, in the Eaton Fire in January 2025. Thankfully the family and goats are safe.
18 McClintock, *Urban Studies and Planning Faculty Publications and Presentations;* Metcalfe, "Let's Goat Crazy." Data on specific urban goat numbers from individual city departments.

Chapter 8: Goat Meat in America

1 "Chevon Official Name Goat Meat," *San Angelo Daily Standard,* June 27, 1922; James Whetlor, *Goat: Cooking and Eating* (Quadrille Press 2018), 13 (epigraph).
2 "Four Storekeepers Fined," *Philadelphia Inquirer,* October 18, 1916; "Goat Meat Sold for Mutton," *New York Times,* August 5, 1899; "To Sell Goat Meat," *Pittsburgh Daily Post,* March 17, 1891; "Goat Meat is Mutton on Ninth Avenue," *The Evening World,* October 26, 1918.
3 William L. Black goes into great detail about the early importation and spread of Angora goats in America in *A New Industry, or Raising the Angora Goat, and Mohair, for Profit* (Keystone Printing Co., 1900). See also Maurice Shelton, *Angora Goat and Mohair Production* (Anchor Publishing Co., 1995); Israel Diehl, "The Goat," in *Report of the Commissioner of Agriculture for the Year 1863* (Government Printing Office, 1863); James C. Bonner, "The Angora Goat: A

Footnote in Southern Agricultural History," *Agricultural History* 21, no.1 (January 1947):42–46; "James Bolton Davis," in *History of the State Agricultural Society of South Carolina from 1839 to 1845* (State Agricultural and Mechanical Society of South Carolina, 1916), 228.

4 United States Department of Agriculture, Bureau of Animal Industry, *Special Report on the History and Present Condition of the Sheep Industry in the United States* (Government Printing Office, 1892), 913–914. The 1890 US Agricultural Census puts the number of sheep in Texas at 3.4 million; see Department of the Interior, Report of the Statistics of Agriculture in the United States, General Tables, Table 7: Sheep and Wool on Farms By States and Territories, Eleventh Census 1890, 236, https://agcensus.library.cornell.edu/wp-content/uploads/1890a_v5-09.pdf; Paul H. Carlson, *Texas Woolybacks: The Range Sheep and Goat Industry* (Texas A&M University Press, 1982), 134; Paul H. Carlson, "Wool and Mohair Industry," *Handbook of Texas Online*, https://www.tshaonline.org/handbook/entries/wool-and-mohair-industry.

5 *Chicago Live Stock World*, October 8, 1903; "Goat Meat Said to be Good," *Chicago Tribune*, August 6, 1899; Chicago Live Stock World, July 11, 1903; "Move for Cheaper Meat," *Kansas City Star*, March 12, 1920.

6 US Congress, *Congressional Record*, 67th Cong., 1st Session, vol. 61, part 1, 1921, 418 (H . 4136 was referred to the Committee on Agriculture); see also US Congress, *Congressional Record*, 65th Cong., 3rd Session, vol. 57, part 3, 1919, 2721; "Marketing Goat Meat," *Angora and Goat Milk Journal*, September 1920, 11; "Mule, Horse and Goat Meat," *Shelbina Torchlight*, May 30, 1919; "Dipping and Goat Meat Marketing Laws Enacted," *Angora and Goat Milk Journal*, April 1921, 5.

7 "Farm and Ranch," *Austin American-Statesman*, May 15, 1887; United States Department of Agriculture, *Seventeenth Annual Report of the Bureau of Animal Industry for the year 1900* (Government Printing Office 1901).

8 Black, *A New Industry*, 12–14 (story of Black's early years in goat ranching); Wyse, "William L. Black," 73–74.

9 "Goats in Chicago," *Belvidere Standard*, May 9, 1888; "Goat Banquet Planned by Livestock Company," *St. Louis Republic*, October 14, 1902; "Angora Goat in Many Styles Served at Novel Banquet," *St. Louis Republic*, October 30, 1902. Hulit established the short-lived National Goat Dairy Company a few years later (see chapter 4).

10 Wyse, "William L. Black," 74–75; Edith Black Winslow, *In Those Days: Memoirs of Edwards Plateau* (The Naylor Co., 1950), 40. See also Carlson, *Texas Woolybacks*, chapter 8, for a detailed account of William Black's Angora ranching years and an account of the canning debacle.

11 Black, *Manual of Angora Goat Raising*, discussion of Angora goat meat beginning at 154; Black's unpublished autobiography is more candid; in it he describes the ranching business as a "comedy of failures," see Wyse, "William L Black," 40.

12 "Chevon Official Name Goat Meat," *San Angelo Evening Standard*, June 27, 1922; "Getting the Goat on Society," *St. Louis Daily Globe-Democrat*, November 5, 1924; "To Name Goat Meat," *Pacific Dairy Review*, June 1, 1922, 10; "B. M. Halbert Takes Issue with Senator E. M. Davis over Chevon," *San Angelo Morning Times*, December 25, 1935; *Pittsburgh Post-Gazette*, June 30, 1922.

13 Abigail Brown Tompkins, letter re: Daniel Tompkins, collection New Jersey Historical Society.

14 "Le Capre Sono Un Cespite Di Lucro," *L'Italia* (San Francisco), April 4, 1918; Ad, *Il Risorgimento Italiano Nel Maryland*, March 1, 1930 (Capretti). Translations by Milena Anfosso. The history of goat entrepreneurship among immigrant communities in the United States during the nineteenth and early twentieth centuries deserves more extensive study.

15 Angora Goat Raisers Face Food Selling Problem, The San Angelo Weekly Standard, February 5, 1971; Federal Register, March 23, 1971, 5435.

16 William M. Grimes, "Label Issue Stirs the Goat Raisers," New York Times, April 11, 1971; "Storm Started by Goat Meat in Hot Dogs," Daily Oklahoman, April 15, 1971; "Hot Dogs? How About Goat Dogs?," The Morning Call (Allentown, PA), April 22, 1971; Federal Register, August 18, 1971, 15740. See also 9 CFR 317.8 (29), the current USDA regulation regarding goat meat labeling.

17 United States Department of Agriculture, Agricultural Statistics Board, National Agricultural Statistics Service, "Sheep and Goats," January 31, 2025, https://downloads.usda.library.cornell.edu/usda-esmis/files/000000018/zk51xc07n/9593wq66x/shep0125.pdf.

18 "Global Snapshot: Goatmeat," fact sheet produced by the trade group Meat and Livestock Australia May 2023, https://www.mla.com.au/contentassets/58fe74b6d47d476189836862e8fe46df/2023-mla-ms_global-goatmeat_final.pdf

19 "How is the Goat Industry Growing," National Animal Health Monitoring System, June 2020, https://www.aphis.usda.gov/livestock-poultry-disease/nahms/studies.

20 United States Department of Agriculture, Agricultural Statistics Board, National Agricultural Statistics Service, "Cattle," January 31. 2025, https://downloads.usda.library.cornell.edu/usda-esmis/files/h702q636h/sf26b275x/h989sz55j/catl0125.pdf; United States Department of Agriculture, Sheep and Goats, January 31, 2025. For MUMS Act, see Public Law 108–282, August 2, 2004.

21 Leslie Svacina, interview by author, March 5, 2020.

22 Yi Yang, "Somali Focus Group Report," https://goats.extension.org/somali-focus-group-report/. See also David Trecheter and Denise Parks, "Somali Goat Meat Preference Survey Fall 2004," University of Wisconsin-River Falls Extension, Survey Research Center Report 2005/3, January 2005.

23 Ayman Noureldin, interview by author, March 11, 2020; "Halal + Kosher Minnesota Meat Market Assessment," University of Minnesota Extension, January 8, 2020, https://conservancy.umn.edu/server/api/core/bitstreams/37ebbc31-3402-4bdb-8511-a21ded9c6bd0/content; Shelby Lindrud, "Halal-certified Goat Processing Facility Coming to Willmar," *West Central Tribune*, March 29, 2023; Christopher Vondracek, "Federal Funds help Launch Halal Goat Meat Slaughterhouse in Central Minnesota," *Minneapolis Star Tribune*, July 12, 2024; Greg Wierschke, interview by author, May 10, 2023.

24 Coleen Taylor Sen, *Curry: A Global History* (Reaktion Books, 2009), see introduction and chapter 1 for a broad discussion of the origins and migration of curry spices throughout the world. Juan Garcia, interview by author, June 17, 2023; Yanikie Tucker, personal communication.

25 Erika Knight, "Evaluation of Consumer Preferences Regarding Goat Meat in Florida" (master's thesis, University of Florida, 2005); Kelyn Jacques, "Midwesterners' Consumer Preference for Goat Meat in a Blind Sensory Analysis" (master's thesis, Oklahoma State University, 2017), 12.

26 "Niman Ranch Founder Starts New Endeavor: Goat Ranching," The Takeaway, WNYC radio, October 15, 2008, interview with Kim Severson, https://www.wnycstudios.org/podcasts/takeaway/segments/8208-niman-ranch-founder-starts-new-endeavor-goat-ranching. Among the articles predicting an upsurge in goat meat consumption are Julie Kendrick, "Goat Meat Could Save Our Food System, But We're Too Afraid to Eat It," *Huffington Post*, October 10, 2018, https://www.huffpost.com/entry/goat-meat_n_5bb64c71e4b028e1fe3bcfa2; A. C. Shilton, "Give Goat Meat a Try, *Outside Magazine* January 24, 2019, https://www.outsideonline.com/food/why-you-should-try-goat-meat/

27 "Goat Meat: A Healthy Choice?," Prairie View A&M University Cooperative Extension, June 2018, https://www.pvamu.edu/cafnr/wp-content/uploads/sites/27/goatmeat_approved.pdf; Snezana Ivanovic, Ivan Pavlović, and Boris Pisino, "The Quality of Goat Meat and its Impact on Human Health," *Biotechnology in Animal Husbandry* 32, no. 2 (January 2016): 117. Goat meat nutritional data is available on the USDA website.

28 Karen Tajonar, Carlos Antonio López Díaz, Luis Enrique Sánchez Ibarra, et al, "A Brief Update on the Challenges and Prospects for Goat Production in Mexico," *Animals* 12 (2022): 837; Pratap Pragna Surinder S Chauhan, Veerasamy Sejian, et al., "Climate Change and Goat Production: Enteric Methane and its Mitigation," *Animals* 8 (2018): 235.

29 Hanna Tuomisto, and M. Joost Teixeira de Mattos, "Environmental Impacts of Cultured Meat Production." *Environmental Science and Technology* 45, no. 14 (2011): 6117–6123; Derrick Risner, Yoonbin Kim, Cuong Nguyen, et. al. "Environmental Impacts of Cultured Meat: A Cradle-to-Gate Life Cycle Assessment," https://doi.org/10.1101/2023.04.21.537778 (UC Davis study); Daniel L. Rosenfeld and A. Janet Tomiyama, "Would You Eat a Burger Made in a Petri Dish? Why People Feel Disgusted by Cultured Meat," *Journal of Environmental Psychology* 80 (April 2022): 101758.

30 Henry Alford, "How I Learned to Love Goat Meat," *New York Times*, March 31, 2009.

Selected Bibliography

Archival Collections
Bancroft Library, Berkeley, California
    Charles Frederick Fisk Papers, 1863–1960
    Chez Panisse Menu Collection
Navajo Nation Library, Window Rock, Arizona
    Office of Navajo Economic Opportunity (ONEO) Oral History Collection
New Jersey Historical Society, Newark, New Jersey
    Abigail Brown Tompkins Papers
San Francisco Public Library, San Francisco, California
    San Francisco History Center Collections
Stanford University Library, Palo Alto, California
    California Historical Society Collection, James Rolph Jr. Papers
University of California, Davis Department of Special Collections
    Charles L. Sullivan Papers on California Wine History
    Edwin C. Voorhies Papers
    Joyce Goldstein Papers on California Cuisine

Government Publications
Hall, S. A., and C. A. Phillips. "Manufacture of Roquefort Cheese From Goat's Milk." University of California College of Agriculture Bulletin 397. University of California Printing Office, 1925.

Diehl, Israel. "The Goat," in *Report of the Commissioner of Agriculture for the Year 1863*. Government Printing Office, 1863.

National Agriculture Statistics Service, US Department of Agriculture, 2017 Census of Agriculture, Navajo Nation Profile.

Navajo Tribal Council Resolutions, 1922–1951. US Government Printing Office 1952.

Hearing Before the Select Committee on Indian Affairs, US Senate, 96th Congress, Second Session on PL 93–531, Report and Plan of the Navajo Hopi Relocation Commission. May 20, 1981.

"How is the Goat Industry Growing?" National Animal Health Monitoring System, June 2020. https://www.aphis.usda.gov/sites/default/files/goat2019-infographic-overview.pdf.

Thompson, George Fayette. *Information Concerning the Milch Goats*. Government Printing Office, 1905.

———. *The Angora Goat*.: Government Printing Office 1901.

Tuberculosis. USDA, Farmers Bulletin 473. Government Printing Office, 1911.

United States Department of Agriculture, Agricultural Statistics Board, National Agricultural Statistics Service, "Sheep and Goats,.," January 31, 2025. https://downloads.usda.library.cornell.edu/usda-esmis/files/000000018/zk51xc07n/9593wq66x/shep0125.pdf

United States Department of Agriculture, *Seventeenth Annual Report of the Bureau of Animal Industry for the Year 1900*. Government Printing Office 1901.

United States Department of Agriculture, Bureau of Animal Industry. *Special Report on the History and Present Condition of the Sheep Industry in the United States*. Government Printing Office, 1892.

United States Public Health Service. *What You Should Know About Grade A Milk*. Public Health Service Publication 1472. US Department of Health, Education and Welfare, 1966.

United States Senate, Committee on Indian Affairs. *Navajos in Arizona and New Mexico*. Part 18. 71st Congress, 3rd Session (1932).

Books, Theses, and Dissertations

Anderson, Virginia DeJohn. *Creatures of Empire: How Domestic Animals Transformed Early America*. Oxford University Press, 2006.

Bailey, Garrick, and Roberta Glenn Bailey. *A History of the Navajos: The Reservation Years*. School of American Research Press, 1986.

Barayon, Ramon Sender, ed. *Home Free Home: A History of Two Open Door California Communes*. Calm Unity Press, 2017.

Bates, Barbara. *Bargaining for Life: A Social History of Tuberculosis, 1876–1938*. University of Pennsylvania Press, 1992.

Belasco, Warren. *Appetite for Change: How the Counterculture Took on the Food Industry 1966–1988*. 2nd updated ed. Cornell University Press, 2006.

Biltekoff, Charlotte. *Eating Right in America: The Cultural Politics of Food and Health*. Duke University Press, 2013.

Black, William L. *A New Industry: Raising the Angora Goat, and Mohair, for Profit*. Keystone Printing Co., 1900.

Brace, Charles Loring. *The Dangerous Classes of New York and Twenty Years Work Among Them*. Wynkoop & Hallenbeck Publishers, 1872.

Brown, Dona. *Back to the Land: The Enduring Dream of Self-Sufficiency in Modern America*. University of Wisconsin Press, 2011.

Brown, Frederick. *The City is More Than Human: An Animal History of Seattle*. University of Washington Press, 2019.

Bull, William Sheldon. *Money in Goats*. The Wakefield Co., 1911.

Cahn, Miles. *The Perils and Pleasures of Domesticating Goat Cheese: Portrait of a Hudson Valley Dairy Goat Farm*. Catskill Press, 2003.

Carlson, Paul H. *Texas Woolybacks: The Range Sheep and Goat Industry*. Texas A&M University Press, 1982.

Campbell, James Wade Hadley. "Exploring The Early Navajo Pastoral Landscape: An Archaeological Study of (Peri)Colonial Navajo Pastoralism from the 18th to the 21st Centuries AD." PhD diss., Harvard University, 2021.

Dunmire, William. *New Mexico's Spanish Livestock Heritage: Four Centuries of Animals, Land and People*. University of New Mexico Press, 2013.

DuPuis, E. Melanie. *Nature's Perfect Food: How Milk Became America's Drink*. New York University Press, 2002.
Farrer, John. *A Perfect Description of Virginia*. Peter Force, 1837.
Frisbie, Charlotte, with Tall Woman and assistance from Augusta Sandoval. *Food Sovereignty the Navajo Way: Cooking With Tall Woman*. University of New Mexico Press, 2018.
Frisbie, Charlotte, and David McAllester, eds. *Navajo Blessingway Singer: The Autobiography of Frank Mitchell, 1881–1967*. University of New Mexico Press, 1978.
Goldstein, Joyce. *Inside the California Food Revolution: Thirty Years that Changed Our Culinary Consciousness*. University of California Press, 2013.
Hall, Sir A. Daniel. *The Improvement of Native Agriculture in Relation to Population and Public Health*. Oxford University Press, 1936.
Hartley, Robert. *An Historical, Scientific and Practical Essay on Milk as an Article of Human Sustenance*. Jonathan Leavitt, 1842.
Horton, Lucy. *Country Commune Cooking*. Coward, McCann and Goeghan Inc., 1972.
Iverson, Peter. *Diné: A History of the Navajo*. University of New Mexico Press, 2002.
———. *For Our Navajo People: Diné Letters, Speeches and Petitions, 1900–1960*. University of New Mexico Press, 2002.
Jacques, Kelyn. "Midwesterners' Consumer Preference for Goat Meat in a Blind Sensory Analysis." Master's thesis, Oklahoma State University, 2017.
Jenkins, Steve. *Steve Jenkins Cheese Primer*. Workman Publishing, 1996.
Jensen, Bernard. *Goat's Milk Magic*. Bernard Jensen, 1994.
Johnson, Broderick, and Ruth Roessel, eds. *Navajo Livestock Reduction: A National Disgrace*. Navajo Community College Press, 1978.
Kauffman, Jonathan. *Hippie Food: How Back-to-the-Landers, Longhairs and Revolutionaries Changed the Way We Eat*. HarperCollins, 2018.
Kelley, Klara, and Harris Francis. *A Diné History of Navajoland*. University of Arizona Press, 2019.
Knight, Erika. "Evaluation of Consumer Preferences Regarding Goat Meat in Florida." Master's thesis, University of Florida, 2005.
Land Reform in the Navajo Nation: Possibilities of Renewal For Our People. Diné Policy Institute, Diné College 2017.
Margolis, David. *Change of Partners*. Permanent Press, 1997.
McNamee, Thomas. *Alice Waters and Chez Panisse*. Penguin Press, 2007.
McNeur, Catherine. *Taming Manhattan: Environmental Battles in the Antebellum City*. Harvard University Press, 2014.
Mihesuah, Devon A., and Elizabeth Hoover, eds. *Indigenous Food Sovereignty in the United States: Restoring Cultural Knowledge, Protecting Environments, and Regaining Health*. University of Oklahoma Press, 2019.
Miller, Timothy. *The 60s Communes: Hippies and Beyond*. Syracuse University Press, 1999.
Miner, Sidney H., and George D. Stanton Jr., eds. *The Diary of Thomas Minor, Stonington Connecticut, 1653–1684*. Day Publishing, 1899.
Morgan, Bill, ed. *The Selected Lettes of Allen Ginsberg and Gary Snyder, 1956–1991*. Counterpoint, 2009.

Mullaly, John. *The Milk Trade in New York and Vicinity*. Fowler and Wells Publishers, 1858.
Myrick, David. *San Francisco's Telegraph Hill*. Howell North Books 1972.
Nicholds, Elizabeth. *Thunder Hill*. Doubleday and Co, 1953.
Ott, Katherine. *Fevered Lives: Tuberculosis in American Culture Since 1870*. Harvard University Press, 1996.
Papanikolas, Helen Z. *The Peoples of Utah*. Utah State Historical Society, 1976.
Richards, Irmagarde. *Modern Milk Goats*. J. P. Lippincott Co., 1921.
Robichaud, Andrew. *Animal City: The Domestication of America*. Harvard University Press, 2019.
Rosenau, M. J. *The Milk Question*. Houghton Mifflin and Co., 1912.
Rothman, Sheila M. *Living in the Shadow of Death: Tuberculosis and the Social Experience of Illness in American History*. Johns Hopkins University Press, 1994.
Sundancer, Elaine. *Celery Wine: the Story of a Country Commune*. Community Publications Cooperative, 1973.
Spink, Wesley William. *The Nature of Brucellosis*. University of Minnesota Press, 1956.
Specht, Joshua. *Red Meat Republic: A Hoof to Table History of How Beef Changed America*. Princeton University Press, 2019.
Stories of Traditional Navajo Life and Culture by Twenty-Two Navajo Men and Women. Navajo Community College, 1977.
Surls, Rachel, and Judith Gerber. *From Cows to Concrete: The Rise and Fall of Farming in Los Angeles*. Angel City Press, 2016.
Thomas, Sherry, and Jeanne Tetrault. *Country Women: A Handbook for the New Farmer*. Anchor Press, 1976.
Van der Donck, Adrian. *Description of the New Netherlands*. Translated by. Hon. Jeremiah Johnson. Directors of the Old South Work, 1896.
Waters, Alice. *Coming to My Senses: The Making of a Counterculture Cook*. Clarkson Potter, 2017.
Weisiger, Marsha. *Dreaming of Sheep in Navajo Country*. University of Washington Press, 2011.
Winslow, Edith Black. *In Those Days: Memoirs of Edwards Plateau*. The Naylor Co., 1950.
Wyse, Christina Reh. "William Black and the Southwestern Livestock Industry." Master's thesis, Texas Tech University 1995.
Young, Alexander. *Chronicles of the First Planters of the Colony of Massachusetts Bay from 1623 to 1636*. Charles C. Little and James Brown, 1846.

Articles

"An Invaluable Milk Supply." *Journal of the American Medical Association*, March 5, 1892.
"Consumption in America." *Atlantic Magazine*, January 1869.
"Erosion in Wakamba Land." *Indians at Work*, March 1, 1934, 13.
"Goat Meat: A Healthy Choice?," Prairie View A&M University Cooperative Extension, June 2018.
"Intravenous Injection of Milk." *British Medical Journal*, August 28, 1880.

"Transactions of the New York Academy of Medicine: A Practical Study of Goat's Milk in Infant Feeding as Compared to Cow's Milk." *The American Journal of Obstetrics and Diseases of Women and Children* (June 1913): 1245–1247.

"Treatment of Tuberculosis by Goat's Serum." *Lancet*, August 24, 1901.

Anderson, Virginia DeJohn. "Thomas Minor's World: Agrarian Life in Seventeenth-Century New England." *Agricultural History* (Fall 2008): 496–519.

Bailey, Flora T. "Navaho Foods and Cooking Methods." *American Anthropologist* (April–June 1940): 270–290.

Bell, J. Finley, M.D. "Some Fat Problems and Goat's Milk in Infant Feeding." *Archives of Pediatrics* 23 (1906): 204.

Bezos, Javier, Lucía de Juan, Beatriz Romero, et al. "Experimental Infection with *Mycobacterium Caprae* in Goats and Evaluation of Immunological Status in Tuberculosis and Paratuberculosis Co-Infected Animals." *Veterinary Immunology and Immunopathology* 133, no. 2–4 (February 2010): 269–275.

Blecha, Jennifer. "Regulating Backyard Slaughter: Strategies and Gaps in Municipal Livestock Ordinances." *Journal of Agriculture, Food Systems and Community Development* 6, no. 1 (Fall 2015): 33–48.

Brinkley, Catherine, and Domenic Vittello. "From Farm to Nuisance: Animal Agriculture and the Rise of Planning Regulation." *Journal of Planning History* 12, no. 2 (May 2014): 113–135.

Bunnell, Mrs. D. L. "California's Women Goat Ranchers." *University of California Journal of Agriculture* (February 1920), 9ff.

Butler, William Hale. "Welcoming Animals Back to the City: Navigating Tensions of Urban Livestock Through Municipal Ordinances." *Journal of Agriculture, Food Systems and Community Development* 2, no. 2 (2012): 1–23.

Butzer, Karl. "Cattle and Sheep from Old to New Spain: Historical Antecedents." *Annals of the Association of American Geographers* 78, no. 1 (1988): 29–56.

Chartier, Craig S. "Plymouth Colony Livestock 1620–1692." http://www.plymoutharch.com/wpcontent/uploads/2010/12/Plymouth_Colony_Livestock.pdf

Daniel, Thomas M. "The History of Tuberculosis." *Respiratory Medicine* 100 (2006): 1862–1870.

Davis, Diana. "Potential Forests: Degradation Narratives, Science, and Environmental Policy in Protectorate Morocco, 1912–1956." *Environmental History* 10, no. 2 (April 2005): 211–238.

Glatman-Freedman, Aharona, and Arturo Casadevall. "Serum Therapy for Tuberculosis Revisited: Reappraisal of the Role of Antibody-Mediated Immunity Against *Mycobacterium Tuberculosis*." *Clinical Microbiology Review* 11, no. 3 (July 1996): 514–532.

Haenlein, G. F. W. "Dairy Goat Industry of the United States." *Journal of Dairy Science* 64 (1981): 1288–1304.

Ivanovic, Snezana, Ivan Pavlović, and Boris Pisinov. "The Quality of Goat Meat and its Impact on Human Health." *Biotechnology in Animal Husbandry* 32, no. 2 (January 2016): 111–122.

Kapture, Judy. "Dairy Goat Industry is Expanding." *Animal Industry Today* (1978), 2ff.

Masumoto, Nancy. "The Nanny Diaries: From Coach Leather to Coach Farm." *Edible Manhattan*, March–April 2012.

Matteson, C. Kieko. "'Bad Citizens' with 'Murderous Teeth': Goats into Frenchmen, 1789–1827." *Journal of the Western Society for French History* 34 (2006): 147–161.

McClintock, Nathan, Esperanza Pollanna, and Heather Wooten. "Urban Livestock Ownership, Management and Regulation in the United States: An Exploratory Survey and Research Agenda." *Urban Studies Planning Faculty Publications and Presentations* 90 (2014): 426–440.

McNutt, S. H., and Paul Perwin. "Tuberculosis of Goats." *Journal of the American Veterinary Medical Association* 12, no.1 (April 1921): 82–84.

Meikle, Virginia, Ann-Kristin Mossberg, Avishek Mitra, et al. "A Protein Complex from Human Milk Enhances the Activity of Antibiotics and Drugs against Mycobacterium Tuberculosis." *Antimicrobial Agents and Chemotherapy* 63, no. 2 (February 2019): aac.01846–18. https://doi.org/10.1128/aac.01846–18.

Oberman, H. A. "Early History of Blood Substitutes: Transfusions of Milk." *Transfusion* 9, no. 2 (March 1969): 74–77.

Olmstead, Alan, and Paul W. Rhode. "An Impossible Undertaking: The Eradication of Bovine Tuberculosis in the United States." *Journal of Economic History* 64, no. 3 (February 2004): 734–772.

Palmer, Mitchell, and W. Ray Waters. "Bovine Tuberculosis and the Establishment of an Eradication Program in the United States: Role of Veterinarians." *Veterinary Medicine International* (2011): 1–12.

Pease, Arthur Stanley. "Some Remarks on the Diagnosis and Treatment of Tuberculosis in Antiquity." *Isis* 31, no 2 (April 1940): 380–393.

Pragna, Pratap, Surinder S Chauhan, Veerasamy Sejian, et al. "Climate Change and Goat Production: Enteric Methane and its Mitigation." *Animals* 8, no. 12 (2018): 235. https://doi.org/10.3390/ani8120235.

Ravenel, Mazyck P. "The Intercommunicability of Human and Bovine Tuberculosis." *Journal of Comparative Pathology and Therapeutics* 112 (1902): 112–143.

———. "Tuberculosis and the Milk Supply." *Journal of Comparative Medicine and Veterinary Archives* 18, no. 12 (December 1897): 753–761.

Riva, Michele A. "From Milk to Rifampicin and Back Again: History of Failures and Successes in the Treatment for Tuberculosis." *The Journal of Antibiotics* 67 (2014): 661–665.

Rodríguez, Sabrina, Javier Bezos, Beatriz Romero, et al., "*Mycobacterium Caprae* Infections in Livestock and Wildlife, Spain." *Emerging Infectious Diseases* 17, no. 3 (March 2011): 532–535.

Rosenkrantz, Barbara Gutmann. "The Trouble with Bovine Tuberculosis." *Bulletin of the History of Medicine* 59, No 2 (Summer 1985): 155–175.

Siddle, David. "Goats, Marginality and the 'Dangerous Other.'" *Environment and History* 15, no. 4 (November 2009): 521–536.

Trecheter, David, and Denise Parks. "Somali Goat Meat Preference Survey Fall 2004." University of Wisconsin-River Falls Extension, Survey Research Center Report 2005/3, January 2005.

Tuomisto, Hanna L., and M. Joost Teixeira de Mattos. "Environmental Impacts of Cultured Meat Production." *Environmental Science and Technology* 45, no. 14 (2011): 6117–6123.

Weisiger, Marsha. "The Origins of Navajo Pastoralism." *Journal of the Southwest* 46, no. 2 (Summer 2004): 253–282.

Wiley, Harvey W. "The Eternal Infantile." *Good Housekeeping*, June 1916.

Yang, Yi. "Somali Focus Group Report." Ohio Cooperative Development Center, 2019. https://goats.extension.org/somali-focus-group-report/.

Index

Chapter note locators are indicated by an *n*: 191n12. Illustrations are indicated by italic locators: *25*.

#GoatSongs, 154

4-H, 96, 116
70 Jamaican Recipes: From My Grandmother's Kitchen (Tucker), 179

A New Industry, Or Raising the Angora Goat, and Mohair, for Profit (Black), 68, 166–167
ADGA (American Dairy Goat Association), 75, 115
African immigrants, 172, 175
Agawam, 80
Agricultural Census of 1900, United States, 37–38
Agricultural Marketing Service, USDA, 182–183
Agricultural Research Center, Beltsville (USDA), 69–70, *69*
agriculture, urban, 28, 148, 149
agritourism, 77, 136, 154
AIDS, 195n14
Allen, Helen, 120, 121
Alpine goats, 47, 102, 104, 113
Alpure brand, 84
Alta Dena Dairy, 108, 115–116
Altadena (CA), 157
alt-casein dairy sector, 135–136
Altham, Emmanuel, 11
American Boer Goat Association, 182
American Boy, The, 75
American Dairy Goat Association (ADGA), 75, 113
American Dairy Goat Record Association, 78
American Goat Society, 201n1
American Kiko Goat Association, 182
American Milk Goat Record Association (AMGRA), 75, 89, 169, 198n4
American Place, An, 128
American Treasure cheeses, 123
AMGRA (American Milk Goat Record Association), 75, 89, 169, 198n4

Ammirati, Giovanni, 89, 169
Anacortes (WA), 79
Andreoli, Lorenzo, 89
Angora Goat Show, 198n7
Angora goats: breeding of, 80; cheese from, 89; land clearing by, 75; and meat industry, 161–171; and Navajo Nation, 34, 47; in Oregon, *163*; at World's Fair, 198n7. *See also* mohair
"Angora Venison," 167
animal concerns: control of, 19–21, 40–41, 144; cruelty to, 43; health of, 174; husbandry of, 72–73; welfare of, 149–150
Anti-Goat Protective Association, 19
anti-goat sentiment, 18–19, 40–42
Apaches de Nabaju, 31
Arapawa goats, 10
Argentina, 86
Arizona, 164
Armour, A. Watson, 81
Armour and Company, 165
artisan cheesemakers, 120, 121, 126–127, 133, 136
Ashe, Elizabeth, 25
Asian immigrants, 175
Atlantic Monthly, The, 54
Australian goat meat, 172–173

back-to-the-land movement: and goat keeping, 95–102; and goat popularity, 113; and milk contamination, 76–77; and new generation cheesemakers, 124, 125; and raw goat's milk, 114; women and, 85
Backus, Barbara, 119
Bacon, Kevin, 154
BAI (United States Bureau of Animal Industry): and animal diseases, 67–68; and breeding, 72; and infant feeding, 70; mutton, use of term, 164; sheep industry survey, 162–163; and tuberculosis, 58–59; and World's Fair, 74–75
Bakewell, Robert, 72
Baldwin Park Goat Cheese Factory, 79
Ball, Gordon, 99–100
Baltimore, 16–17, 146, 160
Baltimore Sun, The, 19
Bamboozle cheese, 136
Baradat, Dr. (French physician), 60
Bare Feet and Good Things to Eat (Boots), 104
Barnes, Almont, 68

Bass Lake Cheese Company, 125
Bates, Caroline, 119
Bauder, Grace, 101
Beard, James, 119
beef production, 173–175
beekeeping, urban, 149, 151
Begay, Aretta, 30
Begay, Nanabah, 35
Bell, Finley, 60
Belle Chevre (Fromagerie Belle Chevre), 129
Beltsville Agricultural Research Center (USDA), 69–70, *69*
Berkshire Blue cheese, 123, 205n8
Bertin, Georges, 60
Bettinehoeve BV, 133
Beyond Meat, 181
BIA (Bureau of Indian Affairs), 33–34, 38–40, 41–45
Bice, Cynthia and Kenneth, 95–96
Bice, Jennifer, 105–106, 116–117, 127, 132
Billy Goat Tavern (Lincoln Tavern), 143
Billydoe Meats, 176
biodiversity, 181
bioidenticals, 134–136
birria (goat meat), 177
Black, Edith, 166
Black, William L., 68, 165–167, *167*
Blair, William M., 170
Blakesville Creamery, 133
blood, goat's, 59–61, 195n14
blue cheese, 90–92, *91*, 123, 128, 205n8
Boer goats, 47, 171, *172*
Bootzin, Robert (Gypsy Boots), 104
Borden, Gail, 81
Bored Cow nondairy milk, 135
Borsodi, Ralph, 95
Bosque Redondo Reservation, 32
Boston Globe, The, 76
Brace, Charles Loring, 25–26
Bragg, Paul, 103–104
Branson, Fannie and Jay, 90–91
Bredenbent, Willem, 10
breeding, 72–73, 80, 113, 116, 136
breeds, goat. *See* Angora goats; goat breeds; Nubian goats; Saanen goats
Briar Hills Dairy, 104
Brier Run Creamery, 125, 126
Brighton (CO), 147–148
Brighton City Council, 148
Broadbent, William, 60
Brooklyn Daily Eagle, 62
Brooks, Barbara, 124, 125, 129
brucellosis, 69, 70
Brunswick Farmers Market, 127
Bubel, Nancy, 98
Buckhead neighborhood (Atlanta), 150–151
Bull, W. Sheldon, 76–77

Bureau of Animal Industry, United States (BAI): and animal diseases, 67–68; and breeding, 72; and infant feeding, 70; mutton, use of term, 164; sheep industry survey, 162–163; and tuberculosis, 58–59; at World's Fair, 74–75
Bureau of Indian Affairs (BIA), 33–34, 38–40, 41–45

c
abrito (goat meat), 175, *178*
Cahn, Miles, 123–124
Calanna Specialty Meats, 171–172
Caldwell, Gianaclis, 151
California: agriculture, urban, 146–147; Angora goats in, 162, 169; cheesemaking in, 87–92, *91*, 95, 119–123, 127–128; cooperatives in, 108, 110–112; counterculture in, 100; early goat population, 18; goat dairy industry in, 77–87, 125, 130–131; health food movement and, 105–106; and Italian immigrants, 24; and Navajo goats, 36; raw milk in, 115–116; and tuberculosis, 65; urban goats in, modern, 149–150, 157–158; women of goat dairy industry, 84–87
"California cliché," 121
California Goat Dairymen's Association (CGDA), 110–112, 203n27
California Milch Goat Association, 78
California Swiss Goat Dairy, 80
Cambridge (MA), 14–15
Camembert, 90
canned milk, 81–84
Canyon Goat Ranch, 81, 87
Cape Verde Islands, 10
Capriblue cheese, 128
Capricese cheese, 128
Capricorn brand cheese, 124
Capriole Farms, 124, 128–129
carbon pollution, 180–182
Caribbean, 9, 178, 179
Carl Hagenbeck's Animal Show, 198n7
Carnation Milk Company, 82
Carpenter, Novella, 149–150
Carroll, Ricky, 126–127
Carson, Rachel, 97
Cary, Kathy, 129
Casa Desierto Rest Camp, 65
casein protein, 134–136
Catholic Worker House, 124–125
Central California Goat Breeders Association, 87
certified milk, 57
CGDA (California Goat Dairymen's Association), 110–112, 203n27
Chaleix, Marie-Claude, 123
Changing Woman, 29–30
Channel Islands, 36

Chatterjee, Choi, 157
cheddar cheese, 108, 109, 112, 124
cheese, goat: artisan cheesemakers, 120, 121, 126–127, 133, 136; in California, 77–87, 125, 130–131; commercial production, 77, 79, 125–130; as health food, 134–136; Indiana cheesemakers, 124; Iowa cheesemakers, 124; Maine cheesemakers, 124, 129, 137, 154; marketing of, 122; Minnesota cheesemakers, 124–125; and Navajo Nation, 35, 37, 45–46; in New England colonies, 13; New Jersey cheesemakers, 126; New York cheesemakers, 124, 126, 136; Northeast production, 123–125; Oregon cheesemakers, 136; Pennsylvania cheesemakers, 136; from raw milk, 108, 137; twentieth century, early, 87–92; Vermont cheesemakers, 124, 126; West Virginia cheesemakers, 125; in Wisconsin, 106–108, *109*, 124–125, 129–133, *132*; Wyoming cheesemakers, 137. *See also* chevre cheese; goat dairy industry; *individual types*
Cheese Board, 120
Cheese Primer (Jenkins), 123
cheese types: Bamboozle, 136; Berkshire Blue, 123, 205n8; blue, 90–92, *91*, 123, 128, 205n8; Camembert, 90; Capriblue, 128; Capricese, 128; cheddar, 108, 109, 112, 124; Chevreese, 126; Emmental, 120; Eurisco, 88, 91; "fauxmage" chevre, 134; feta, 108, 109, 123; gjetost, 77; "goatless," 134; gouda, 123; Jack, 119, 120; Lutzelfluh Emmental, 120; mozzarella, 88, 108; Neufchâtel, 106; nondairy, 134–136; organic, 125; ricotta, 88, 109; Roquefort, 90–92, *91*; Sonoma Jack, 120; Wensleydale, 128
Cheesemakers' Journal, 127
chefs, American, 119–120, 127–129, 180
Chenel, Laura, *118*, 119–121, *122*–123, 125, 127–128, 129–130
chevon (goat meat), 161, 167, 168, 170, 184
chevre cheese, 119–123, 126, 127–128, 129, 134, 205n8. *See also* cheese, goat; Chenel, Laura; goat dairy industry
Chevreese cheese, 126
Chevrock Farm, 91
Chez Panisse, *118*, 119–121, 180
Chicago: animal control efforts, 149; and dairy distilleries, 52; and goat dairy industry, 71, 80–81; and goat meat marketing, 165–166; and infant feeding, 66; and labeling scandals, 161, 163; slaughterhouses in, 165; and tuberculosis, 63–64, *64*; and urban goats, 143, 149, 151–152
Chicago Cubs, 143

Chicago Livestock World, 163
chickens, urban: in California, 157; and communal living, 99, 106; as gateway animal, 87, 141, 151; in Los Angeles, 79; in New York City, 23; and ordinances, city, 144, 148–149; in Portland, OR, 156; survey of, 159
Children's Aid Society, 25–26
Chilton Dairy, 131, *132*
Chinese immigrants, 189–190n26
Church, Genevieve, 158
Church of the Lukumi Babalu Aye, 150
Church of the Lukumi Babalu Aye, Inc. vs. City of Hialeah, 150
Churro sheep, 46
cider, hard, 16, 188n12
City Grazing, 158, *159*
city ordinances. *See* ordinances, city
civil rights movement, 96
Clah, David, 44
Clapp, Roger, 13
climate change, 46, 180–182
Coach Farm, 123–124
Coburn, Loren, 83
co-grazing, 181
Cohill, Edmund P., 75
Coit, Henry, 57, 116
Collier, John, 42
"Collier goats," 43
colonial era, 10–13
Columella (Roman writer), 56
Committee on Indian Affairs, Senate, 39
Common Market, 106
communes, counterculture, 96–100, 106, 117
community building, 156–158, 159
Compton (CA), 146
Compton, Griffith, 146–147
condensation, milk, 81–84
Conlin, Richard, 141–142, 158
Connecticutt Colony, 12
consumption. *See* tuberculosis
control, animal, 19–21, 40–41, 144
Cook County Goat Commission, 63–64, 80
Cooperative League of the United States, 105
cooperatives, food, 105, 127
cooperatives, goat dairy, 108–112, 119, 203n20, 203n21, 203n24
Cosenza, Millanella, 28, 190n32
Cotswold sheep, 33
"cottage food" laws, 137
Council, Jennifer, 147–148
Counselor, Jim, 43
counterculture, 95–102
Country Commune Cooking (Horton), 99
Country Journal of Vegetable Gardening, The (Bubel), 98
Country Women: A Handbook for the New Farmer (Country Women collective), 102

Country Women collective and commune, 102
Country Women (magazine), 102
Cove Mountain Goat Dairy, 115
Covid-19, 154, 156, 174, 183
Cowgirl Creamery, 206n18
cow's milk: cheese, 87, 136; condensation of (evaporation), 81–82, 110; contaminated, 3, 56–57, 61–62, 63, 73; distillery dairies, 51–53; Grade A milk regulations and, 114; and immigrant cheese production, 89; and infant feeding, 66–67; popularity of, 51; and poverty, 15; and safety laws, 136; and tuberculosis, 53–57, 61–62, 63, 65, 168, 194–195n8
Crampton Goat Dairy, 78
creation story, Diné/Navajo, 29–30
cuisine, American, 119–120, 121
cuisine, Caribbean, 178
cultured meat products, 181
curd, frozen, 131–133, 206n20
curry goat, 178–179
Cylon Rolling Acres, 175
Cypress Grove Cheese Company (Cypress Grove Chevre), 87–88, 102, 128, 129, 130, 132
Cypress Island (WA), 79

Daily Alta California, 18
Daily Oklahoman, 170
dairy, cow. *See* cow's milk
dairy, goat. *See* cheese, goat; chevre cheese; goat dairy industry
dairy breeds, 130–131
Dairy Goat Journal, 102–103, 126, 132
Dairy Goat Princess of California, 117
Dairy Goat Products Cooperative, 203n21
Davis, Adelle, 99
Davis, Gary R., 195n14
Davis, Heather, 155
Davis, James B., 162
Dean and DeLuca, 127
Deardourf, Rush, 102–103
Denetsosie, Hoke, 48
Denver Goat Dairy, 78
Department of Animal Science, UC Davis, 90
destructive habits, 14–15, 16–17, 18–19, 40–41. *See also* ordinances, city; urban goats, historical; urban goats, modern
Detroit, 148–149
Detroit City Council, 149
Detroit Tigers, 143
Diné Community Advocacy Alliance, 46
Diné people. *See* Navajo Nation
Diné Policy Institute, 46
Diné/Navajo creation story, 29–30
Dinétah (ancestral home of Navajo), 33

direct marketing, 175
disease, germ theory of, 58
Diseases of the Stomach and Bowels of Cattle (BAI Circular 68), 68
distillery dairies, 51–53, 56
distributor safety documentation, 136
Dobbs, Floyd, 108
Donck, Adrian van der, 12
drought, 37–38, 46
Drumlin Dairy, 131
Duncan, Penny, 124, 127
Dutch Hill (NYC), 22–24, 23, 25–26

East River Industrial School, 25–26
Edgar, Gordon, 134
Edwards Plateau, 165
Eggers, Melvin (M. P.), 77, 104
Eid al-Adha (holiday), 175–176
El Chivar Ranch, 65, 80–81, 87, 89
Elawa Farm, 81
"Emergency Goat and Kid Meat Biscuit," 169
Emmental cheese, 120
Emmi Group, 129, 130, 133, 206n18
Encaria Goat Farm, 77
encomienda (forced labor), 31
entrepreneurs, goat dairy, 78–81
environmental degradation, 38–40
"Equal-To" Act (Wholesome Meat Act), 169–170
ethnic markets, 177
ethnic prejudice, 24–25, 32
Ettersburg (CA), 88
Ettersburg cheese factory, 91–92
Eurisco cheese, 88, 91
European dairy goats, 73–74, 94, 167–168
Evanston Goat Milk Company, 64
Evanston Hospital, 64
evaporation, milk, 82–84
events, goat-themed, 154–156
Evers, Medgar, 96

Fairway Market, 123
Fantome Farm, 127
Farlowe, W. E., 166
Farm City: The Education of an Urban Farmer (Carpenter), 149
Farm commune, The, 97, 99
farm stores, urban, 153
farmers markets, 127, 175
farming, regenerative, 175
farmstead dairy production, 136
"fauxmage" chevre cheese, 134
Fawsitt, Cormac B., 49
FDA (United States Food and Drug Administration), 66, 174
feedlots, 173–174

Fellowship Farm, 106–107
Fermify, 135
Fernández, Antonio, 43
Fernández, Juan, 9–10
Ferris, Nancy Pierson, 98
feta cheese, 108, 109, 123
fiber/textiles. *See* mohair; wool
Finland, 57–58
Fish Market, 128
Five Points, 24
Fletcher, Janet, 121
Flight from the City (Borsodi), 95
Food & Wine (Monthly Magazine of Food and Wine), 122
Food and Drug Administration, United States (FDA), 66
food cooperatives, 105, 127
"Food For Thought" (Mayer), 115
Food Freedom Act (WY), 137
Food Institute, 134
Food Safety Act, 169
food safety laws, 136, 169
Foot Rot in Sheep: Its Nature, Cause and Treatment (BAI Bulletin 63), 68
forced labor (*encomienda*), 31
forced migration, 32–33
Ford, Henry, 195n14
Fore Street, 129
Forgoine, Larry, 128
Fort Amsterdam, 11
Fort McKavett Tanning Company, 165
Fort Wayne Weekly Sentinel, 76
Fort Wingate (NM), 34
Foster's Freeze, 111
4-H, 96, 116
Fourant, Denise, 126
Fraga Farm, 136
Frank Leslie's Illustrated Newspaper, 53
freedom of religion, 150
French, James E., 88
French Alpine goats, 73
French cuisine, 122, 128
French-style goat cheese. *See* chevre cheese
Frisco Livestock Company, 166
Fromagerie Belle Chevre (Belle Chevre), 129
frozen curd, 131–133, 206n20
fruit trees, 15–16

ganado menor (small livestock), 164
Garcia, Juan, 177, *178*
Gaskin, Stephen, 97
gateway animals, 149, 151. *See also* chickens, urban
Georgetown University, 70
Gerber, Karl, 120
germ theory of disease, 58
German immigrants, 22, 23

"Get Your Goat" (Pierson), 98
Getting Back Together (Houriet), 96–97
Gies, Michael, *101*
gjetost cheese, 77
GlennArt Farm, 151
globalization of goat dairy industry, 129–130
"Goat Babies in Pajamas" video, 154, *155*
goat breeds: Alpine, 47, 102, 104, 113; Arapawa, 10; Boer, 47, 171, *172*; European dairy, 73–74, 94, 167–168; French Alpine, 73; Kiko, 171, 182; LaMancha, 119; miniature, 141–142; Navajo, 35–37, *37*, 38–40; Nigerian Dwarf, 150–151, 153, 156; Pygmy, 144, 157; Saanen-Toggenburg, 151; Savanna, 171; Schwarzenburg, 73; Schwarzwald Alpine, 198n7. *See also* Nubian goats; Saanen goats; Toggenburg goats
goat cheese salad (Chez Panisse), 121
goat dairy industry: antitrust lawsuits, 112; breeding practices, 72–73; in California, 77–87, 130–131; challenges to, 134–136; chefs/restaurants promoting products, 127–129; commercial production, 125–130; condensed/evaporated milk production, 81–84; economics of, 35–36; emergence of, 77–81; equipment and supplies, 125, 130; European dairy goats, 73–75, 167–168; farmstead production, 136–137, *137*; frozen curd in, 131–133, 206n20; globalization of US, 129–130; and goat meat, 171; growth of, late century, 112–113; industrial production, 130–138; local milk supply for, 133; and low-income population, 110; marketing, 80–81, 83, 89, 91–92, 94, 103, 110–111; mid-century industry, 93, 122–125; organizations, professional, 74–75; price manipulation, 133; publications for, 76; raw milk (unpasteurized), 108, 113–116; resurgence of interest in, 113; supply, milk, 108; and tuberculosis, 58–67; UC Davis research on, 89–90; in Wisconsin, 106–108, 124–125, 129–133, *132*. *See also* cheese, goat; chevre cheese; cooperatives, goat dairy; milk, goat's
Goat Hill (NYC), 22–26, *23*
Goat Justice League, 142
Goat Lady of Potrero Hill, 143–144
Goat Mafia (food truck), 177, *178*
goat meat. *See* meat, goat
"goat meat panic," 161–164
"Goat Meat Sold for Mutton" (*NYT*), 161
Goat Milk Journal, 85, 102–103
Goat Milk Producers Cooperative, 109
Goat Ordinance of 1920, 28
goat panic, 39–40
goat reduction, the (historical event), 42–44
Goat Rodeo Farm, 136

Goat Works, 126, 127
Goat World, The: and cheesemaking, 87; and Florian Heintz, 78–79; on goat dairy industry, 93; and goat enthusiasts, 76; on goat meat-mutton debate, 168; on Melvin Eggers, 77; on milk production, 69; Navajo goats advertisement in, 37; on Stevens auction, 81; on tuberculosis, 59
goat yoga, 136, 151, 154–155, *156*
"goatless" cheese, 134
goat's blood, 59–61, 195n14
Goat's Milk Magic (Jensen), 105
"Goats Yelling Like Humans" video, 154
#GoatSongs, 154
Goattober, 183
Gold Rush, 17–18, 67, 147
"Golden Immigration," 18
Goldman, Bobby, 108
Goldstein, Joyce, 120
Good Housekeeping, 66
Gottlieb, Lou, 100
gouda cheese, 123
gourmet food, 127, 180
Gourmet magazine, 119
grade A milk, 114, 117
Grade A Pasteurized Milk Ordinance (PMO), 114
Grant, Jennie, 141–142, 158
grazing: co-grazing, 181; control of, 40–41; multi-species, 181; overgrazing, 31, 38–40; permit system, 43, 44–45; urban, 21, 22, 24, 158, *159*
Great Depression, 92, 171
Greatorex, David, 128
Greek immigrants, 143, 169, 176
Green, Gael, 121
Greystone Chevratel, 128
Griffith, Alice, 25
grocery stores, general, 110–111
Guadalupe Island, 36
Gypsy Boots (Bootzin, Robert), 104

Habig, Frank, 77–78
Haedtler Goat Milk Dairy Farm, 78
Hagenbeck, Carl, 198n7
Halal Farms USA, 176
halal processing, 175–176
Hall, Alfred Daniel, Sir, 41
Hall, Bolton, 76
Hall, Hope, 154
Halona, W. Mike, 45
Hamor, Ralph, 11, 12
Hankin, Michael, 106–108, *107*
Hanna, Pam, 100
Hannaford grocery, 127
Happy Halal, 176
hard cider, 16, 188n12

Hardgrave, Mrs. E. W., 161
Hartley, Robert, 53
Harvard University, 56
Hatch Act (1887), 68
Hathaway, Margaret, 156
"Have-More" Plan: A Little Land—a Lot of Living, The (Robinson), 95, 98
Hayden, Carl, 164
Hayward, Sam, 129
health, animal, 79, 174
health food movement, 103–106, 113–115, 117
health food stores, 105, 127
Health Hut, 104
Healthy Diné Nation Act, 46
heat treatment. *See* pasteurization
Hedrich, Clara and Larry, 171–172
Heintz, Florian F. T., 78–79, 80–81
Helvetia Milk Condensing Company, 81–82
Hendrick the Tailor, 10–11
Hialeah (FL), 150
Hidden Valley Health Ranch, 104
Higginson, Francis, 11
hippies. *See* communes, counterculture
Hippocrates, 54
Hispanic immigrants, 175
Historical, Scientific and Practical Essay on Milk, An (Hartley), 53
Hmong food, 177
Holistic Goat Care (Caldwell), 151
Holsum Dairies, 131
Home Free Home (Morningstar Ranch), 100
homesteading, 95, 98, 125, 141. *See also* back-to-the-land movement
hoof-and-mouth disease, 79
Hooper, Allison, 124, 126
Hopi Reservation, 39
Horton, Lucy, 98–99
Houriet, Robert, 96–97, 99
House of Representatives, United States, 164
Howell, Joseph, Jr., 39
Howland, Martha, 87, 89
Howland, Winthrop, 65, 80–81, 89
hózhó (state of balance), 30, 48
Hulit, A. B., 80, 166
Humboldt Times, 88
husbandry, animal, 72–73
Hwéeldi (place of suffering), 32–33

Illinois, 64, 151, 176
immigrants: Caribbean, 9, 178, 179; cheesemakers, 88–89; and Dutch Hill, 23; goat dairy entrepreneurs, 89; and goat meat, 175–176; Greek, 37, 143, 169; Hmong, 177; Indian, 176; Irish, 22, 24–25; Italian, 28, 37, 88–89, 169; keepers, 22, 24–25, 28, 37, 168–170, 183; Pakistani, 176, 177;

and Potrero Hill, 143; Somali, 176; and Telegraph Hill, 21
immunity, tuberculosis, 58–62, 70
imported goat milk, 131–133
Impossible Foods, 181
Improvement of Native Agriculture in Relation to Population and Public Health, The (Hall), 41
Indian Health Service, 62–63
Indian immigrants, 176
Indian Removal Act, 32
Indian Territory, 32, 33
Indiana cheesemakers, 124
Indiana Colony, 65
Indians at Work (BIA), 41
Industrial Revolution, 17, 51
infant feeding, 66–67, 70, 77
Information Concerning The Angora Goat (Thompson), 68
Information Concerning the Milch Goats (Thompson), 72, 73
INStyle magazine, 154
internet marketing, 175
internment, Navajo, 32–33
Ioder, Carolyn and David, 151, *152*
Iowa cheesemakers, 124
Irish immigrants, 22, 24–25
Italian immigrants, 24, 88–89, 169

Jack cheese, 119–120
Jackson, Andrew, 32
Jackson-Mitchell Company, 112, 129, 203n27
Jalisco (Mexico), 147, 177
Jamaica, 178
Jamestown, 11
Jams, 128
Jasper Hill Dairy, 133
Jenkins, Steve, 123, 127
Jensen, Bernard, 104–105
Jeter, John, 112
Jewel Grocery, 127
Jochemsen, Andries, 10
Johnson, Edward, 11–12
Johnson, Lyndon B., 169
Johnson, Samuel, 10
Josselyn, John, 12
Judge (New York), 25
Judgement of Paris, 120
junk food, 46

Kady, Roy, 47–48
Kafka, Barbara, 127
Kansas City (MO), 164–165
Kapowsin Dairy, 128
Keats, John, 54

Keehn, Mary, 101–102, 128
Keeping Goats for Profit (Barnes), 68
kefir, 106
Kehler, Mateo, 133
Kenya Agricultural Commission, 41
Kerrville Sanitarium, 65
Kickapoo of Wisconsin cheddar, 108
Kiko goats, 171, 182
Kilmoyer, Bob and Letty, 129
King, Martin Luther, Jr., 96
Kirby, Violet, 81, 87
kneel down bread, 35
Koch, Robert, 54–55, 70
Krivit, Karen, 158

labor/mining camps, 169
LaClare Family Creamery, 131, 171–172
LaDuke, Winona, 46
LaMancha goats, 119
Lamb Marketing Board, 183
land management, sustainable, 158, 159
Land o'Lakes Inc., 129, 130
landscape deterioration, 46
land-use regulations, 144–146, 148
Larrupin Café, 128
Las Cabritas Farm, 85, *85*
Las Vegas Hot Springs, 65
Laura Chenel Company, 129–130, 132, 206n20
Laurelwood Acres Goat Dairy, 116, 123
Le Gourmand, 128
Leach, Corl A., 103
Leach, Corl Eber (C. E.), 102–103
Leary, Timothy, 96
LeCompte, Gail, 126, 127
Leicester sheep, 72
Les Copains, 128
Let's Cook it Right (Davis), 99
Lily's Bistro, 129
Lincoln Tavern (Billy Goat Tavern), 143
Lindo, Jake, 20
L'Italia (newspaper), 83, 169
Little Rainbow Chevre, 123
Lively Run Goat Dairy, 136
livestock, Southwest, 30, 33–34, 37–38, 39, 42–44, 48
Livestock Congress Hall (St. Louis World's Fair), 75
livestock regulations, 141–142
Living the Good Life (Nearing), 95
Livingston, Denisa, 46
Lone Oak Goat Farm, 78
Long Walk, the, 32, 43
Lopez, Raymond, 149
Los Angeles Times, 80–81, 105
Los Angeles urban agriculture, 146–147
Louisiana Purchase Exposition, 74

low-income population, 110
Lullaby brand, 84
Lutzelfluh Emmental cheese, 120

M. tuberculosis (mycobacterium tuberculosis). See tuberculosis
Madison (WI), 106–108, 127
Maefsky, Christine and Vincent, 124–125
Magic Forest Farm, 100
Maher, James, 17
Maine cheesemakers, 124, 129, 137, 154
Malta fever, 69
Manheim Township (PA), 144
Manufacture of Roquefort Cheese from Goat's Milk (UC Davis), 90
Margolis, Judith, 99–100
maritime distribution of goats, 9–10
marketing: of cheese, goat, 91–92; of dairy goats, 80–81, 89; direct internet, 175; of evaporated goat's milk, 83, 110; by French government, 122; health food, goat milk as, 103; of meat, goat, 165–166, *167*, 168–169, 175, 182–184; of plant-based products, 134–135; and tuberculosis, 94
marketing boards (USDA), 182–183
Marshall, James, 17–18
Martyn Sanitorium, 65
Mary's Infant Asylum and Maternity Hospital, St., 67
Massachusetts Bay Colony, 11–12
Mayer, Jean, 115
McDonald, Peter, 45
meat, goat: Angora goats, 161–169; availability to consumer, 177; canned, 165; in colonial era, 12–14; and communal living, 99–100; cuisines, in ethnic, 177–179; deceptive labeling, 161–162, 163–164; domestic demand, 172; early industry, 161–171; economics of production, 173–175; in gourmet cuisine, 180; health of animals, 174; healthful properties, 180; during holidays, 175, 176; immigrant demand for, 175–176; imported, 172–173; and Italian immigrants, 24, 89; marketing of, 165–166, *167*, 168–169, 175, 182–184; in Minnesota, 175; modern industry, 171–179; mutton, sold as, 161–164, 168–170; and Navajo Nation, 34–35, 46–47; processing, commercial, 174–176; processing of urban animals, 149–150; taste for, popular, 179–180, 184; from urban goats, 147, 149; in Wisconsin, 175. *See also* chevon (goat meat)
meat, plant-based, 181
meat, sheep, 161
meat alternatives, 181
meat industry, 164–165

mega-dairies, 131
Melody Acres neighborhood (Tarzana, CA), 146
Mendel, Gregor, 72–73
Merino sheep, 33, 47
methane production, 180–181
Mexican cuisine, 177
Mexican-American War, 32
Meyenberg, John B., 81–82, 84
Meyenberg, John P., 84
Meyenberg, Walter, 82, 83
Meyenberg Company, 110, *111*, 112, 129, 203n24
Meyenberg Old-Fashioned Products, 111
Miami, 150
Michael Reese Hospital, 63, 66–67
Michael's Restaurant, 128
migration, forced, 32–33
milk, cow's. *See* cow's milk
milk, goat's: and communal living, 99–101; demand in twentieth century, 35–36; evaporated, 82–84, *82*, 110; government interest in, 67–70, *69*; as health food, 103–106, 113–115, 195n14; and infant feeding, 66–67, 70, 77; as medicine, 58–67; and Navajo Nation, 34–35, 37, 45–46, 47; in New England colonies, 13, 14, 15; raw (unpasteurized), 104, 108, 113–116, 147, 151; transfusions of, 60; and tuberculosis, 58–67; ultra-pasteurized, 112; waning interest in, 201n1. *See also* cheese, goat; chevre cheese; goat dairy industry
milk, plant-based, 134–136, *135*, 181, 195n14
Milk Goat in California, The (Voorhies), 90
Milk Goat Journal, The, 76, 102
Milk Source company, 131
Miller, George, 82
miniature goats, 141–142
Minnesota cheesemakers, 124–125
Minor, Thomas, 13–14
Minor Use and Minor Species Animal Health Act (MUMS Act), 174
Miracle Brand evaporated milk, 110, 112
Mitchell, Frank, 35
Modern Farmer, 155
Modern Milk Goats (Richards), 85–86, *86*
mohair: and Bureau of Animal Investigation, 68; during Great Depression, 171; industry in American south, 162–163; and Italian immigrants, 89, 169; and Navajo Nation, 34, 40, 47
Money in Goats (Bull), 77
Monroe (OR), 154–155
Montchalin, Michael, 92
Montchevre brand, 129, 133
Monthly Magazine of Food and Wine (*Food & Wine*), 122

Monticello Farmers Mutual Co-op Creamery, 109
Moosetrack brand cheese, 124
Morning Call, 170
Morningstar commune, 97
Morocco, 41
Morris, Edmund, 95
Morrow, J. W., 79
Morse, Lainey, 154–155
Mother Earth News, 97, 98, 125–126
Mount Taylor, 33
mozzarella cheese, 88, 108
Ms. magazine, 102
Mt. Lassen Goat Dairy, 92
Mt. Sterling Co-op Creamery, 108, *109*
multi-species grazing, 181
MUMS Act (Minor Use and Minor Species Animal Health Act), 174
Mungo, Raymond, 100
municipal zoning, 27
Murphy the Goat, 143
Murray Hill, 25–26
Muslim immigrants, 175, 176
mutton, goat meet as, 161–164, 168–170
mycobacterium tuberculosis (M. tuberculosis). See tuberculosis
Mycobacterium types, 56, 70
Mystic Lakes Goat Dairy, 115, 204n31

Naftaly, Bruce, 128
Nassaikas, Nicholas J., 88
National Animal Monitoring System, 173
National Food Shop, 105
National Goat Dairy Company, 80
National Tuberculosis Eradication Program, United States, 58, 93
Navajo Blessingway Singer (Mitchell), 35
Navajo Churro sheep, 47
Navajo Department of Natural Resources, 45
Navajo goats, 35–40, *37*
Navajo Lifeway Inc., 30
Navajo Livestock & Trading Company, 36
Navajo Nation: ancestral home of, 33; anti-goat sentiment and, 41–42; creation story, 29–30, 190n1; and drought, 37–38; economy of, 43; environmental degradation in, 38–40; food systems of, 45–46; goat breed, 30, 35–40; livestock in, 30, 31, 33–34, 42–44; origins of, 30–31; shepherding in, 44–45, 46–47; United States and, 32–34, 48
Navajo National Trust Land Leasing Act of 2000, 45
Navajo Times, 44, 45
Navajo Tribal Council, 42, 44, 45
Navajo weavers, 47
Nearing, Helen and Scott, 95

Neighbors Opposed to Backyard Slaughter, 150
Nelson, Ernest, 43
Neufchâtel cheese, 106
New Amsterdam, 10–11
New England Cheesemaking Supply Company, 126–127
New England colonies, 10–13
New Haven (CT), 15
New Jersey cheesemakers, 126
New Mexico, 30, 31
New Mexico Territory, 32
New Seed Starters Handbook, The (Bubel), 98
New York Agricultural Experiment Station, 67
New York cheesemakers, 124, 126, 136
New York City: distillary dairies in, 53; Dutch/Goat Hill neighborhood, 22–24; goat cheese promotion in, 119–120, 123, 127–128; and goat meat, 169; and immigrant keepers, 26–27; and nuisance goats, 19, *20*, 26–27; Union Square Green Market, 127
New York City Common Council, 19
New York Magazine, 121
New York Observer, 53
New York Times, The, 60, 96, 121, 142, 161
Newbold, Douglass, 128
Newmark, J. P., 19
Nicoletti, Luigi, 88–89
Nigerian Dwarf goats, 150–151, 153, 156
Niman Ranch, 180
nineteenth century, 16–26, 28
No Goats, No Glory, 142
Nocard, Edmond, 59
nondairy cheese, 134–136
Noreldin, Ayman, 176
Norfeldt, Wes, 116, 123
North Clayton Cheese Factory, 108
North Counties Milk Goat Association, 88
Northeast goat cheese, 123–125, 126
Northern Navajo Agency, 34
"Notes by a New Goat Keeper" (Bubel), 98
Novak, Albert, 63
Nubian goats: cheese at restaurants, 128; on Cypress Island, 79; and infant feeding, 66–67; at Little Rainbow Chevre, 123; and Navajo Nation, 47; and Roquefort cheese, 91–92
nuisance laws. *See* ordinances, city
Nūmi, 135

Oak Forest Tubercular Hospital, 63, *64*
Oakland (CA), 149–150
Oakland Planning Commission, 150
Oakville Grocery, 127
obesity, 46

Ode to Goats (Fawsitt), 49
Oklahoma City Council, 115
Old Goat Hater, 39
Oñate, Juan de, 30
OPEC oil embargo, 113
ordinances, city: Brighton (CO), 148; cultural concerns, 150; Detroit (MI), 148–149; historical, 11, 14–15, 18–19; in San Francisco, 28; in Seattle, 141–142
Oregon, 136, 162, *163*, 164
Oregon Dairy Goat cooperative, 110
Oregon Territory, 67
Oregon Trail, 67
organic agriculture/food, 97–98, 105, 115, 125, 136, 180, 203n20
Organic Gardening (*Rodale's Organic Gardening and Farming*), 97–98
Organic Valley Cooperative, 203n20
organizations, professional dairy, 74–75
orishas, 150
Oscar Mayer brand, 165
Osteopathic Health Resort, 65–66
Ovens of Brittany, 107
Oxford Down Sheep Association, 198n4
Ozark Dairy Goats Products Cooperative, 109

Pacific Coast Condensed Milk Company, 82
Packers Corners (Total Loss Farm), 100, *101*
pajarete (beverage), 147
Pakistani immigrants, 176, 177
parasites, 173
Parker, Hy, 79
Parks, Rosa, 96
Parnell, Liz, 129
Pasadena (CA), 65
pasteurization, 56–57, 113–115, 137, 204n29
Patten, Amanda Louise Buchanan, 63–64, 80, 81
Patten, James, 63–64
Pederaza, Veronica, 133
Peet, Alfred, 119–120
Peluso Cheese Company, 112
Pennsylvania cheesemakers, 136
People magazine, 122
Pepin, Jacques, 128
Perfect Day, 135
Perfect Description of Virginia, A, 11
PET milk company, 82
pets, goats as, 150–151
Philadelphia: goat cheese in restaurants of, 128; and goat meat labeling, 162; goat nuisance, 17, 21, 22; ordinances, city, 19; urban goats in, 27; urban goats in, modern, 158
Philadelphia World's Fair 1876, 74
Philly Goat Project, 158

Phoenix Academy of Cultural Exploration and Design, 106
phthisis. *See* tuberculosis
Picq, Jules, 60
Piedmont Sanitarium (VA), 63
Pike, Arthur E., 65
Pingree Potato Patch Program, 148
Piper, Kathleen, 106–107
Pitts Goat Dairy, 78
Pixton, Alfred E., 88
plant-based "dairy," 134–136, *135*, 181, 195n14
plant-based "meat," 181
Plymouth colony, 11
Polk, James, 162
Polled Hereford Association, 73
Poplar Hill Dairy, 125
Portland (OR), 156–157
Potato Famine, 24
potbellied pigs, 150, 156
Potrero Hill (San Francisco), 27, 28, 143–144, *145*
poverty, goats and, 15, 20–27, 24, 64, 72, 113, 163
prejudice, ethnic/racial, 24–25, 32, 39–40, 63
Prohibition, 79
Prudhomme, Paul, 119–120
Public Health Service, United States, 114
Puck (New York), 25
Puck, Wolfgang, 128
Pueblos, 30–31
Pygmy goats, 144, 157

Q fever, 141

Radnor-Wayne Sanatorium, 65
Rainbow Chevre, 205n8
Rambouillet sheep, 33, 47
Range Canning Company, 165, *167*
rangeland goats (Australia), 173
raw milk (unpasteurized), 104, 108, 113–116, 147, 151
Redwood Empire Dairy Goat Association, 116–117
Redwood Hill Dairy and Creamery, 129, 133
\ Hill Farm, 103, 105–106, 117, 127, 132
Redwood Hill Farm Goat Dairy, 96
Reed, Barbara, 123
Reese, Bob, 124
regenerative farming, 175
regulatory requirements, 136
"remystification of food," 134
Republic, The, 17
Rians Group, 129, 130
Richards, Irmagarde, 83, 84–87

Richland Farms neighborhood (Compton), 146, 147
ricotta cheese, 88, 109
Robinson, Carolyn and Ed, 95, 98
Roby, Lelia Foster (Mrs. Roby), 71–72, *71*, 74–75
Rodale's Organic Gardening and Farming (Organic Gardening), 97–98
Rogue Gold Dairy, 110
Rolph, John, 27–28
Roquefort cheese, 90–92, *91*
Rose Parade, 78
Roth Kase, 206n18

Saanen goats: on Cypress Island, 79; Irmagarde Richards and, 86; at Laurelwood Acres Goat Dairy, 116; as pure European breed, 77; research, and USDA, 69, 70; and Roquefort cheese, 91–92
Saanen-Toggenburg goats, 151
Sage, Lily, 158
Salmon, Daniel E., 68–69, 163
salmonella, 149
Salt Lake City, 88–89
samp (porridge), 13
San Francisco: animal control in, 20, 25, 27–28; Board of Health, 190n32; Chinese immigrants in, 189–190n26; and Cosenza Millanella, 190n32; and evaporated milk, 83, 85; gourmet food shops in, 120; immigrant entrepreneurs in, 89, 169; and meat, goat, 169; and nuisance, goat, 16, 143–144, *145*; poor neighborhoods, early, 21–22, *22*, 24–25, 27–28; stray livestock in, 18–19; and urban grazing, 158, *159*
San Francisco Board of Health, 190n32
San Francisco Board of Supervisors, 28
San Francisco Examiner, 144, 190n32
San Francisco General Hospital, 66, 85
San Haven Sanitorium, 65
San Joaquin County Medical Commission, 116
San-a-Tog Dairy, 77
sanitariums, tuberculosis, 62–66, *64*, 79, 94, 195n14
Santa Barbara General Hospital, 66
Santa Fe (NM), 31
Santeria, 150
Saputo Inc., 129, 132, 133, 206n18
Saunders, Rose, 81, 87
Sava, Verena, 125, 126
Savanna goats, 171
Saxelby, Anne, 183
"Say Fromage" promotion, 122
"Say No to Frozen Curd" campaign, 133
Sayeed, Omer, 157
Sayer, Jere Linda, 124

Sayer, Phil, 124
Schack, Steve, 117
Schad, Judy, 124, 128–129
Schulman, Phyllis, 142
Schwarzenburg goats, 73
Schwarzwald Alpine goats, 198n7
screaming goat meme, 154
Seal Cove Farm, 124, 125, 129
Seattle Department of Planning and Development, 141
Secor, William, 65
Sedgewick, Kyra, 154
Senate Committee on Indian Affairs, United States, 39
Serum (podcast), 195n14
Seven Cities of Cibola, 30
seventeenth century, 13–16
Severson, Kim, 180
sexism in farming, 85–86
Shafor, William A., 73, 75, 198n4
Shanty Town (NYC), 26
sheep: Churro, 46; Cotswold, 33; and drought, 38–39; Leicester, 72; meat of, 161; Merino, 33, 47; Navajo Churro sheep, 47; Rambouillet, 33, 47; in Texas, 167–168, 170
Sheep is Life organization, 45, 47
shepherding, 29, 44–48
Sheraton, Mimi, 121
Sherman, Dick, 109
Sianis, William "Billy Goat," 143
Silent Spring (Carson), 97
small-farm homes, urban, 147
Smith, Henry, 106
Smith, Joanna Guthrie (née Joanna Guthrie), 106, 107
Smith, Virgil D., 39
Smitt, Claes Claesen, 10
social media, 154
social reform, 25–26
social welfare, 156–158, *159*
Somali immigrants, 175, 176
Somes, Erica, 156–157
Sonoma Cheese Factory, 119
Sonoma County, 95
Sonoma County Fair, 117
Sonoma County Goat's Milk Cheese, *118*
Sonoma Jack cheese, 120
Sonoma Mission Creamery, 112
South Austin neighborhood (Chicago), 151
Southeast Asian immigrants, 172
Southern California Goat Association, 87
Southwestern Range Sheep and Breeding Laboratory, 34
Southwestern Wisconsin Dairy Goat Products Cooperative, 108, 124
Spago Restaurant, 128
Spanish colonists, 30, 31, 178

Special Milk Products, 112
St. Louis World's Fair 1904, 74–75, 198n7
St. Mary's Infant Asylum and Maternity Hospital, 67
Standardized Milk Ordinance, 114
Star Spangled Foods, 127
Starbucks Coffee, 120
Steamboat Rock Creamery, 109
Stege, Camilla, 124
Stetson, Bob and Debby, 129
Steuve brothers, 115
Steve Allen Show, The, 104
Stevens, Charles A., 80
Stonington (CT), 13
streptomycin, 55, 94
Sunflower Farm Creamery, 154, *155*
Supreme Court, United States, 150
sustainable land management, 158, 175
Svacina, Leslie, 175
Swiss Goat Dairy, 77–78

Tarzana (CA), 146
Tate, James, 149
Teal, Albert, 90–91
Telegraph Hill Neighborhood Center, 25
Telegraph Hill (San Fran), 21–22, *22*, 24–25, 28
temperance movement, 53
Ten Acres Enough (Morris), 95
Ten Apple Farm, 156
Texas, 29, 162, 165–168, 170
Texas Sheep and Goat Raisers Association, 167–168, 170
textiles/fiber. *See* mohair; wool
The American Standard Milch Goat Keeper, 76
Thompson, George Fayette, 67–69, 72–75, 198n4, 198n7
Thoreau, Henry David, 54
Three Acres and Liberty (Hall), 76
Time magazine, 142
Toggenburg goats: and Beltsville Agricultural Research Center (USDA), 69, 70; at Canyon Ranch, 87; on Cypress Island, 79; as European breed, 73; at GlennArt Farm, 151; and Irmagarde Richards, 86; at Las Cabritas Farm, 85; at Little Rainbow Chevre, 123; marketing of, 77; and Navajo Nation, 34, 35–36; and Roquefort cheese, 91–92, *91*; at St. Louis World's Fair, 75; Stevens' auction of, 80–81; and tuberculosis, 63, *64*, 65–66; UC Davis research on, 89–90
Toggenburg-Saanen goats, 151
Tomales Bay Foods, 206n18
Tomorrow Farms (Tomorrow Foods), 134–135
Tompkins, Daniel F., 169
Tompkins, David, 74–75

Topham, Anne, 107, 127
Total Loss Farm (Packers Corners), 100, *101*
transfusions, goat's milk, 59–60
tuberculin, 55
tuberculosis: and cow's milk, 53–57, 61–62, 63, 65, 168, 194–195n8; decline of, 93–94; and goat's milk, 103–104, 114
Tuberculosis Eradication Program, United States National, 58, 93
Tucker, Yanikie, 179

UC Davis Department of Animal Science, 90
UC Davis (University of California, Davis), 89–90, *91*
Ulhorn, Bill and Nancy, 110
Union Square Green Market, 127
United States Agricultural Census of 1900, 37–38
United States Bureau of Animal Industry (BAI): and animal diseases, 67–68; and breeding, 72; and infant feeding, 70; sheep industry survey, 162–164; and tuberculosis, 58–59; at World's Fair, 74–75
United States Department of Agriculture (USDA), 57–58, 67–70, 90, 168, 170–171, 176
United States Department of Agriculture's National Tuberculosis Eradication Program, 93
United States Food and Drug Administration (FDA), 66, 174
United States Government and Navajo Nation, 32–34, 48
United States House of Representatives, 164
United States National Tuberculosis Eradication Program, 58, 93
United States Public Health Service, 114
United States Senate Committee on Indian Affairs, 39
United States Supreme Court, 150
University Farm School (CA), 89–90
University of California, Davis (UC Davis), 89–90, *91*
University of Minnesota, 176
urban agriculture, 28, 147, 148, 149
urban goats, historical: dwindling of, 26–27; New York City, 22–27, *23*; nuisance of, 10–11, 14–17, 18–26, 62; opposition to, 143–144; Philadelphia, 27; poverty indicator, as, 24; reform of, 25–26; San Francisco, 21–22, *22*, 24–25, 27–28
urban goats, modern: challenges of keeping, 151–154; community, creating, 156–158; and cultural differences, 150; and farm stores, 153; home processing of meat, 149–150; and land-use regulations, 144–146, 148; in Los Angeles, 146–147;

opposition to, 147–148; public interactions with, 153; reasons for keeping, 150–151; scope of keeping, 158–160; and social media, 154. *See also* chickens, urban; ordinances, city
urban planning, 17, 27, 143–144
USDA (United States Department of Agriculture), 57–58, 67–70, 90, 168, 170–171, 176
USDA Agricultural Marketing Service, 182–183
USDA Agricultural Research Center in Beltsville, MD, 69–70

value-added products, 136
veganism/vegetarianism, 99, 100
Vella family cheesemakers, 110
Vermont Agency of Agriculture, 124
Vermont Butter and Cheese Company (Vermont Creamery), 124, 129, 130
Vermont cheesemakers, 124, 126
Vermont Creamery (Vermont Butter and Cheese Company), 124, 126, 129, 130, 132
"victory buck" program, 94
Vietnam War, 96
Villemin, Jean-Antoine, 58–59
Vineland Acres, 78–79
Voorhies, Edwin, 90

W. G. Tobin's Chili con Carne, 166, *167*
Wagner, Morris, 85
Wall Street Journal, The, 112–113
Wallace, E. R., 115
Wallis, Samuel, 10
Walnut Acres neighborhood (Woodland Hills, CA), 146
Waltham Goat Dairy, 78
Warwick (RI), 15
Washington DC, 17
Washington State, goat dairy in, 77, 79, 91, 115, 128, 162
Waters, Alice, 119–121, 180
Waxman, Jonathan, 128
weavers, Navajo, 47
Wensleydale cheese, 128
West, Estelle, 143–144, *145*
West Virginia cheesemakers, 125
Westfield Farm, 129
Westwyndes Goat Dairy, 78
What You Should Know About Grade A Milk (United States Public Health Service), 114
Wheeler, Burton, 42
Whex, 104–105
White, Jane Storey, 81, 87
White Dog Café, 128
Whole Earth Catalog, 97–98

Wholesome Meat Act ("Equal-To" Act), 169–170, 172
WHYY (Philadelphia), 195n14
Wicks, Judy, 128
Widemann, Alfred, 82
Widemann, Chris (C. H.), 82–84, 110–111
Widemann Goat's Milk Company, 82–84, *82*, 85, 110
Wieninger, Sally and Theodore, 123
Wierschke, Greg, 176
Wiley, Harvey W., 66
Wilson, Carl G., 59
Wine and Cheese Center, 120
Winger Cheese Company, 124
Winslow, Edward, 11
winter cheese production, 132
Winthrop, John, Jr., 12
Wisconsin: and back-to-the-land farmers, 106; and Charles A. Stevens, 80; goat cheese in, 106–108, *109*, 124, 125, 129–131, *132*, 133; goat meat production, 175; population, dairy goat, 29
Wisconsin Dairy Goat Products Cooperative, 124
Wolf, Clark, 127
women of goat dairy industry, 84–87
Wood, William, 13
Woodland Hills (CA), 146
wool, 34, 47, 162–163
Woolwich Dairy, 129
World Series, 143
World's Fair 1904, St. Louis, 74–75, 198n7
Wrigley, William, Jr., 81
Wrigley Stadium, 143
Wyoming cheesemakers, 137

yoga, goat, 136, 151, 154–155, *156*
York Hill Creamery, 124, 127

Zabar's (NYC), 124
Zeh, William, 38
Zeiler, Ian, 126
zoning, municipal, 27, 204n31